T0285480

# Praise for *Return to the Sky*

"Three cheers for this splendid, surprising, inspiring book! First, for its eloquent author, who as an anxious and inexperienced young woman took on a tough project too important to fail. Second, for the eaglets who, thanks to her innovative care, grew up to reclaim American skies. And lastly, a cheer for you, dear reader—for when you close these covers, you'll know you, too, may help repair our beautiful, broken world."

—Sy Montgomery, naturalist; author of *The Soul of an Octopus*

"*Return to the Sky* is a vivid reminder of how the actions of a single person can have the best kind of ripple effects. In this case, that person is Tina Morris, who figured out how to reintroduce Bald Eagles in the Northeast during the 1970s. If you look up and see an eagle almost anywhere in the United States, chances are you should thank this visionary woman."

—Scott Weidensaul, author of *A World on the Wing*

"Tina Morris's trailblazing memoir lifts us to new heights, like one of the Bald Eagles she raised taking a first flight into the wilds. Her full-circle story highlights a lifetime of making this world a better place, from nurturing animals and children to rewilding a home place."

—Marina Richie, author of
2024 John Burroughs Medal–winner *Halcyon Journey*

"In *Return to the Sky*, Tina Morris eloquently recounts how embracing passion and overcoming fear empowered her to undertake an adventure in helping to save the endangered Bald Eagle. It reminds us that every challenge begins with a dive into the unknown, born of conviction and the courage to transcend seemingly insurmountable odds. Along the way, Morris's journey becomes intertwined with the fate of the Bald Eagle at a time when the future of this iconic species teetered on the brink."

—Michael J. Caduto, author of *Pond and Brook* and *Through a Naturalist's Eyes*; coauthor of the *New York Times* best-selling *Keepers of the Earth* series

"Here is the true, compelling story of how one woman with enormous self-determination followed her dream to bring Bald Eagles back to their lost habitats. Tina Morris is one determined, caring biologist who found herself in the right place at the right time to bring eagles back to our skies. She is an inspiration to follow your dreams."

—Stephen W. Kress, coauthor of *Project Puffin*

"In *Return to the Sky*, Tina Morris shares the personal, inside story of what it was like to raise the first Bald Eagle chicks that jumpstarted the recovery of this endangered bird species. Over two summers, Tina served as a surrogate eagle mom, developing methods that would later be used by biologists in ten states and the District of Columbia to raise over a thousand more Bald Eagle chicks for release into the wild. Along the way, she discovered her own self-confidence and direction in life."

—Tom French, PhD, former director of the Massachusetts Natural Heritage and Endangered Species Program

# Return to the Sky

# Return to the Sky

## The Surprising Story of How One Woman and Seven Eaglets Helped Restore the Bald Eagle

Tina Morris

**Chelsea Green Publishing**
White River Junction, Vermont
London, UK

Developmental Editor: Matthew Derr
Project Manager: Natalie Wallace
Copy Editor: Ashley Davila
Proofreader: Nancy A. Crompton
Designer: Melissa Jacobson
Page Layout: Jenna Richardson

Printed in the United States of America.
First printing October 2024.
10 9 8 7 6 5 4 3 2 1    24 25 26 27 28

**Our Commitment to Green Publishing**
Chelsea Green sees publishing as a tool for cultural change and ecological stewardship. We strive to align our book manufacturing practices with our editorial mission and to reduce the impact of our business enterprise in the environment. We print our books on chlorine-free recycled paper, using vegetable-based inks whenever possible. This book may cost slightly more because it was printed on paper that contains recycled fiber, and we hope you'll agree that it's worth it. *Return to the Sky* was printed on paper supplied by Sheridan that is made of recycled materials and other controlled sources.

**Library of Congress Cataloging-in-Publication Data**
Names: Morris, Tina, 1949– author.
Title: Return to the sky : the surprising story of how one woman and seven eaglets helped restore the bald eagle / Tina Morris.
Description: White River Junction, Vermont : Chelsea Green Publishing, [2024]
Identifiers: LCCN 2024024956 (print) | LCCN 2024024957 (ebook) | ISBN 9781645022633 (hardcover) | ISBN 9781645022640 (ebook) | ISBN 9781645022657 (audio)
Subjects: LCSH: Bald eagle—Reintroduction—New York (State) | Women in science—New York (State)—History—20th century.
Classification: LCC QL696.F32 M666 2024 (print) | LCC QL696.F32 (ebook) | DDC 598.9/4209747--dc23/eng/20240705
LC record available at https://lccn.loc.gov/2024024956
LC ebook record available at https://lccn.loc.gov/2024024957

Chelsea Green Publishing
White River Junction, Vermont, USA
London, UK
www.chelseagreen.com

For my family
And all our animals
Past and present

*It is well to fly towards the light,*
*even where there may be*
*some fluttering and bruising of wings . . .*

—Elizabeth Barrett Browning

# Contents

## Part Three: 1977



## Part Four: Beyond Eagles



## Part Five: Eagles Aloft

# Foreword

In the realm of bird conservation, the tale of the Bald Eagle's resurgence stands as a powerful testament to the enduring strength of nature and the impact of dedicated individuals. Return to the Sky by Tina Morris vividly chronicles her journey of determination, ingenuity, and hope that directly contributed to the Bald Eagle's comeback. This book highlights how a single person's unwavering passion and dedication can drive remarkable recovery and restoration for an entire species.

As the CEO of the National Audubon Society, I have the privilege of hearing and witnessing countless stories of individuals committed to protecting our natural world. Tina's journey stands out as particularly inspiring. Her devotion to the Bald Eagle, a species emblematic of successful American wildlife recovery efforts, is extraordinary.

Growing up along the Potomac River, I never saw a Bald Eagle in the wild. However, these majestic birds are commonly sighted along the Potomac today, and they are a beacon of renewal and hope. This transformation did not happen by accident. It is the result of concerted efforts by passionate advocates like Tina, whose work underscores the importance of preserving our natural heritage for future generations.

The revival of the Bald Eagle mirrors Audubon's mission: to protect birds and the places they need, today and tomorrow. Audubon's initiatives align around a focused effort to bend the bird curve, aiming to halt and ultimately reverse the decline of bird populations across the Americas. Tina's dedication to seven eaglets is a shining example of this effort in action. Through her project, she demonstrated how targeted intervention, combined with community support and scientific knowledge, can lead to the recovery of a species on the brink of extinction. Tina's story is evidence of the power of focused conservation efforts in bending the bird curve, showing us that this vital goal is achievable with commitment and collaboration.

*Return to the Sky* also reminds us of the critical role of community and grassroots action in environmental work. Tina's efforts were not just about saving the birds but also about engaging and inspiring a community to rally around a common cause. This aligns with Audubon's belief in the transformative power of individual actions united by a shared purpose. Every person's contribution, no matter how small, can lead to monumental changes when we work together toward a common goal. Tina's story exemplifies how grassroots movements can amplify individual efforts, creating a ripple effect that leads to substantial, long-lasting impacts on conservation and the environment.

As women in science, Tina and I have both navigated a field historically dominated by men. When I entered graduate school a decade after Tina, the field was still predominantly male, but progress had been made thanks to pioneers like her who opened doors for others. Tina's achievements exemplify the invaluable contributions of female conservationists who work to protect our planet, overcoming significant barriers along the way. Her legacy not only celebrates these contributions but also inspires us to continue to strive for greater inclusivity and recognition of diverse voices in our efforts to protect nature.

*Return to the Sky* is more than an account of environmental triumph; it is a call to action. At a time of urgent climate and biodiversity crises, this book challenges each of us to examine our surroundings and consider how we can contribute to the sustainability of our planet. The stakes have never been higher, and our window for meaningful action is rapidly closing.

This book goes beyond recounting the past; it aims to inspire future generations of environmental stewards. Tina and others like her have blazed a trail, proving that one person's steadfast dedication can create historic change. The path she has forged underscores the importance of supporting and empowering the next generation of scientists and conservationists.

Tina's compelling narrative and tireless work serve as a powerful example of what can be achieved through dedication and passion.

# Foreword

As you read this book, I have no doubt you will find inspiration and motivation to contribute to our shared mission of protecting the planet.

It is now our responsibility to carry forward this legacy, ensuring that the skies remain a safe haven for Bald Eagles and all the other magnificent birds that grace our natural world. Together, we must rise to the challenge, taking bold actions to safeguard our environment for generations to come.

Dr. Elizabeth Gray
CEO, National Audubon Society

# Prologue

**June 12, 2015**

On a clear June day in upstate New York, a rabbit dined on the sweet grasses at the edge of a country road, oblivious to its fate. A Bald Eagle soared over an open field searching for his next meal. With his keen vision, as much as eight times stronger than that of the average human, and quick reflexes, the bird spotted the rabbit and dove upon it, yellow talons at the ready. The predator's size and powerful beak made quick work of its defenseless quarry. Distracted by his kill, the eagle did not see the approaching vehicle hugging the shoulder of the road, until it was too late. As the lifeless bird lay next to his prey, the sun reflected off a small aluminum band attached to one of his legs, etched with the numbers 03142.

Such collisions are one of the most common threats to eagles. Less dexterous than falcons or hawks, their large size helps them overcome their prey efficiently but slows their reaction time. In this case, the lives of both predator and prey ended in one final meal. This might have been considered a tragic event. No one wants to see or even read about animals killed by cars, especially an animal as charismatic and symbolic as a Bald Eagle. But for me, the news was a cause for celebration.

I had heard of the eagle's demise in an unexpected way. The school year was nearing its end; the students had left for the day. I was cleaning up my desk, sorting through final papers and exams for my ninth-grade biology class as I considered which to take home to grade that night. While shutting down my laptop, I spotted the press release, dated June 12, 2015, in the mire of my inbox. It looked official, and although I was tempted to leave it for later, the headline caught me—*Oldest Bald*

# Prologue

*Eagle ever in US, 38, found dead in NY.* I scrolled down to read the story, transported back nearly forty years.

> *The state Department of Environmental Conservation . . . responded to a report of a dead eagle in . . . Monroe County, on June 2, after it was hit by a car. According to the banding records, the bird was a nestling brought from . . . northern Minnesota as part of New York's Bald Eagle Restoration Program.*

This eagle was not ordinary. Its leg band, listed in the US Geological Survey (USGS) database, was easy to track. When someone reports a banded bird to a state or federal agency, valuable information about its life story can be revealed. Number 03142, it turns out, had an illustrious history that explained why the press release, issued by the Department of Environmental Conservation (DEC) in Albany, had made its way to me. I never found out who had forwarded it, but someone in the DEC office had made the connection between us.

The news of 03142's death hit me in a personal way. Our lives had been entwined at the start of his life, and even though thirty-eight years had passed, the bond I felt with this bird hadn't weakened with time. I was a part of his story and he of mine. In 1976 and 1977, the eyes of the state and federal government and of everyone hoping to see the national bird restored to the sky were focused on a remote hilltop in central New York's Montezuma National Wildlife Refuge, an hour north of Cornell University, where seven Bald Eaglets, taken from nests in the Midwest, would be the pioneers of a historic initiative designed to reintroduce Bald Eagles to the Northeast. Ultimately, the success or failure of this program fell on the shoulders of an unlikely source. The young, adventurous, but inexperienced and uncommonly anxious graduate student researcher, who was handed the responsibility for this project, tasked with keeping these eagles alive so they could help return their species to the skies, was me.

The eagle's tragic end gave those of us involved in the thirteen years of the Bald Eagle Restoration Project final proof of its success, success embodied in one eagle who had survived, despite the odds, to help repopulate its species.

# Prologue

I owed much to 03142, or M3, as I called him when we first met. My life had been a circular path of self-discovery, fraught with thwarted ambition and dead ends, yet illuminated by occasional truths. In my perennial search for meaning and direction, I had followed many blind leads, never knowing where they would take me. In 1976, I found myself in the right place and time to assume the responsibility for a ground-breaking conservation project, including working with a Bald Eagle whose life had been likewise affected by happenstance. He taught me our futures are often mapped by serendipity rather than prior plans. He, with a broken leg sounding his death knell, and I, with confusion and self-doubt paving my route to adulthood, helped save one another. By giving me purpose, M3 straightened out the detours, twists, and turns of my life. On July 4, 2026, the country will celebrate its 250th birthday while simultaneously honoring the 50th anniversary of the reintroduction of its national bird. I will be celebrating the life of M3.

# Part One

# Finding the Path

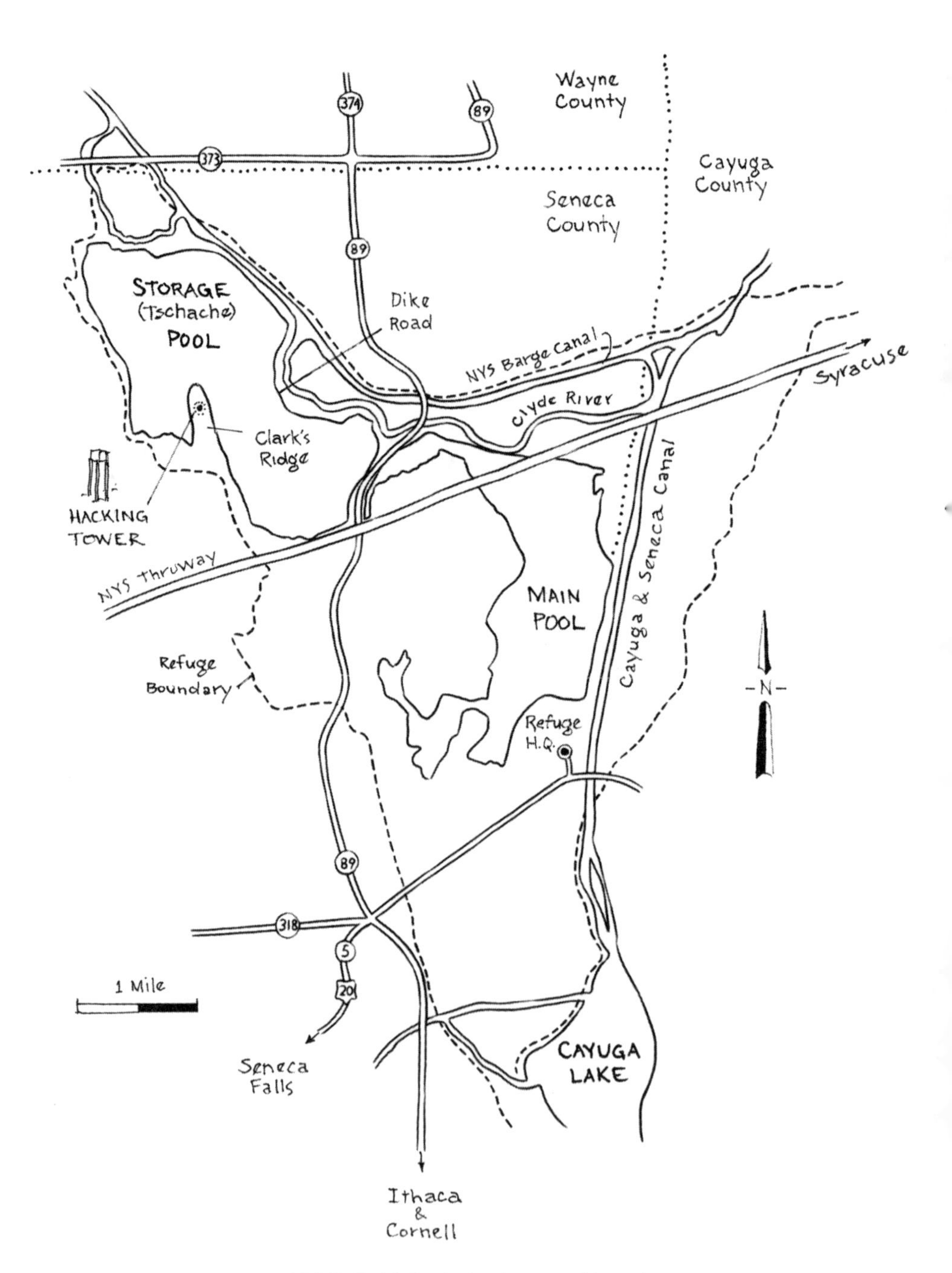


Map of Montezuma NWR Bald Eagle hacking and breeding sites. *Illustration by Michael Waters*

## — Chapter 1 —

# Tails of Childhood

I wasn't an only child who needed company, or a frustrated rebel with parents who refused to allow pets in the house for fear of allergies or dirt, or a collector of exotic critters that I could show off to my friends. Animals simply entranced me. Despite living next to a farm, my family kept its own collection of horses, goats, donkeys, the usual dogs and cats, and assorted orphans in cages. But injured wildlife was my passion. My heart was broken at least once a week when the nest of baby rabbits unearthed by the dogs, or the fledgling robin dislodged by the cats, didn't make it, despite my tender loving care.

Dawn held special significance in my early life. Because my dad loved horses and insisted all six of his children know how to ride, I was in a saddle before I learned how to walk. Every morning before school, the two of us would rise before the sun, mount our horses, and head down the trail in darkness, leaving my mother and siblings to welcome morning in their own time.

As light appeared on the horizon and our pace quickened, beads of sweat formed under the brow of my riding helmet. Then we were off, an easy trot evolving into a canter, its speed depending on the terrain and deviations of the trail. In sections, jumps appeared along the path, usually formed by fallen logs, causing my anticipation to mount and my hands to tremble with dread. As I grew older, the jumps grew in height, logs replaced by three-to-four-foot post and rail fences encircling the horse pastures around us. My goal was to hide my fear from my father, so desperate was I to retain my status as his loyal and stalwart riding companion.

My favorite part of the ride came when we slowed to a lazy walk. Now we could visit the foxes' den where the kits cavorted outside the entrance, and most thrilling of all, could watch hawks circle above the fields searching for movement of prey below them. Sometimes, we would catch an owl perched overhead, preening, and settling in for its daily nap. The splendor and independence of these raptors elevated them to another dimension, beyond that of mammals and other birds. My father didn't know species names or the details of their natural history, but his enthusiasm instilled in me an appreciation of the wild world.

My mother bred golden retrievers, so there were as many as fourteen dogs in the house at one time when there was a litter—five or six adults and the rest puppies. Goldens are a gentle breed but like all dogs that move in a pack, they ravaged any creature they encountered. The half-eaten squirrels, bunnies, chipmunks, and birds they brought back from their forays, most beyond help, went into my animal "hospital," which I had set up in the bathtub. Cardboard boxes lined with cotton held the injured creatures, most of them babies with eyes still closed. Using eye droppers and moistened hardboiled egg yolk or chopped up earthworms, I nursed them until they died. Usually this happened quickly, but a few I was able to save and release back into the woods.

Under a giant beech tree, I built a graveyard for the animals who didn't make it. Here I would grieve for them as I marked their deaths with stick crosses, contemplating which career to embark on when I was grown. Doctor or veterinarian seemed fitting since I loved caring for the sick. At nine, I fretted that if I didn't pick a career, my life would be devoid of meaning.

Once I was set on studying medicine, either human or veterinary, I made the mistake of telling my parents about my ambition. Their enthusiasm made me wonder whether they saw it as possible redemption, since I think they were worrying a bit about their daughter's obsession with nonhuman creatures. Being a doctor (they conveniently ignored the vet option) would somehow save me. Commensurate with my desire to please, my dreams shifted from rescuing small critters to wearing a white coat and stethoscope.

As early as high school, reality dampened my dreams. In biology class, each student was given a cat cadaver to dissect. It reeked of formaldehyde and looked like it had been lost in a torrential rainstorm, wandered into my lab station, and died. Since this was my first experience with a science lab course, I steeled myself and dove into the cat corpse, identifying all the arteries, veins, and organs I could find, fighting my urge to flee.

The next nightmare came when we were told to design and conduct an original experiment. I chose one that tested the need for Vitamin C in guinea pigs. Expected to cause a little weight loss, the experiment went tragically wrong. Two days later, their stiff round bodies lay in the cage, tiny feet aimed skyward and dried blood protruding from their mouths. It was evident animal testing was not for me.

---

My misadventures in science continued to haunt me during my first two years at Barnard College, culminating with my anatomy class. When it came to memorizing bones and systems, I was a star. But when I found out the labs involved testing the endurance of rabbits on a treadmill, I was done. I approached the professor's desk after class and announced I was sorry but would have to be excused from the experiments, because I could not abuse a rabbit. The ensuing conversation altered my academic path. The teacher was sympathetic, which I hadn't expected, but unyielding. The lab portion of the course would be working on animals, he explained, not just rabbits, but frogs and cats as well. I had to complete it or lose credit for the class. Then he asked in a gentle tone, as if talking to a small child, "Why are you taking this course, anyway, since it's optional for the biology major?"

"I want to go to medical school," I replied, "and it's required for pre-med."

"You realize that many of the classes you need to take will involve animals, don't you? You'll be doing dissections and experiments and metabolic studies. How do you plan to handle them? Even vet school requires these labs."

Today, fifty years later, most medical and vet schools have dispensed with live animal labs, bowing to public pressure that questioned their ethics and effectiveness. I was just unlucky to be ahead of the movement for more humane treatment of animals in school. At that moment, I could see I had no business studying medicine. My pre-med career was over; a pre-vet major wasn't an option either. Without a goal to motivate me, my interest in school ebbed, leaving me to hobble my way through the rest of my sophomore year. I yearned for change but was so lost I had no idea where to seek it.

My mother noticed my malaise. Having spent several years living in Germany when she was my age, she encouraged me to do a study abroad program. Her reasoning made sense: while learning a different language and culture, I could figure out what I wanted to do. My father was still hanging on to the idea of having a doctor daughter. After convincing him that this was not just a waste of my time—a "boondoggle" as he called it—I headed to the University of Grenoble in France for my junior year. Classes at the university resembled a mini-United Nations. As foreign students, we all became part of a close-knit expat community, sharing stories about our respective lives over meals or informal gatherings in cafes. My world expanded as I became intrigued by customs and social mores so different from my own.

---

By the time I returned to the States a year later, I had decided an anthropology major would suit me better than science. By studying their culture, I could learn about people's relationships with the animals sharing our Earth. Looking back on my decision, I was grasping for an alternative to learn about wildlife while avoiding the dissections and animal experiments. Perhaps my curiosity about animals and my desire to protect them would have to become an avocation in my life.

My memories of New York City and Barnard, an urban enclave of concrete and congestion, were still raw enough to convince me a major change in geography, as well as size and college culture, was in order. Oberlin College in Ohio, with its intimate campus perched on the southern plains of Lake Erie, boasting fields and trees and even farms, as well as a strong anthropology department, met my requirements. My life had turned upside down—it wouldn't be the last time—but a change in direction might offer a new beginning.

Convinced that science lived in the lab with petri dishes, microscopes, and fleshy specimens inside jars, a realm both confining and sterile, I had abandoned pre-med courses at Barnard. Physics and chemistry defeated me, requiring more focus than I was ready to give them, and biology labs not only didn't connect to my fascination with animals but belied my desire to protect them.

My switch to anthropology at Oberlin opened a gateway to the spiritual connections between humans and animals. To avoid science, I had chosen an alternative pathway into the world of wildlife, but something was still missing. I was hovering above the periphery of my life and needed to find the entrance. It wasn't until the spring of my senior year, while reviewing career possibilities, that my ambition began to crystallize—elusive and undefined yet emerging from the mists of my mind.

In late May of our final semester, a friend invited me to join him on a bird survey he'd been assigned for his ecology class. I didn't know one bird from another unless it was a blue jay or chickadee, but I was game.

As we walked through the woods with binoculars affixed to our eyeballs, he identified every bird from its call, even when we could see only leaves. It was migration season in northeastern Ohio, the epicenter of the Atlantic and Mississippi flyways, so by day's end we had an extensive list of birds, including Scarlet Tanagers, Brown Thrashers, Cedar Waxwings, and Wood Thrushes, all of which I was meeting for the first time. Nowhere did I spot the familiar robin or other common backyard denizens. Instead, we were surrounded by an avian world of color and diversity, strangers all. Dozens of tiny warblers, each looking the same with their yellow bodies and black markings scattered from head to tail, and brown sparrows named for where they lived (field, swamp, savannah, tree) rather than their appearance, were the most baffling. Even when my friend shouted, "Oh, look there, an Indigo Bunting!" it didn't mean anything except I admit to being astounded by its stunning color. I was peering through a kaleidoscope into what had previously been a barren forest scene, introduced to the magical realm of birds for the first time. Captivated by their wildness and winged emancipation from the human world, we stared up at them, our binoculars and bird books in hand. But they had their own agenda, oblivious to our presence, as they pecked frantically at seedpods hanging from branches, sang love songs to attract their mates, uttered aggressive squawks to ward off their competitors, or carried forest detritus to their nests. They were intent on survival, and we were inconsequential.

Although our walk demonstrated to me that science could be practiced outside with wildlife, in its natural habitat, this revelation arrived late. With graduation a month away, my commitment to a degree based on an understanding of culture—a human construct—hadn't prepared me for a career involving animals.

# — Chapter 2 —

# A Year at the Zoo

Three weeks later, in June 1972, I graduated from Oberlin in a bewildered state, not too different from the way I had entered college four years earlier. Few options were open to us. The Vietnam War narrowed the possibilities for my male friends: graduate school, in some cases to escape the draft; community service jobs as conscientious objectors; Canada; or the military and potential deployment to the war. A pall settled over campus that spring before graduation as antiwar sentiment, always strong at Oberlin, permeated the student community. Male students received their draft numbers, and we all worried about the fate of our friends whose numbers were low. For us women, our choices were more open-ended without the draft hovering over us, but in some sense no less difficult to make. While the war didn't call us to the front lines, the call to make some difference in the world was a large part of our education, and we found ourselves coveting the rudders that guided the young men.

I was rescued from needing to make a decision, at least for the short term, by a group of classmates who, by pooling their savings, had purchased a hundred acres south of Louisville, Kentucky, for an organic farm. They had invited me to visit after graduation. My tent was in the trunk, and I was ready to walk the banks of the Nolin River, which bordered their property, while contemplating my next move.

As US Route 42 heads southwest, the rolling hills of southern Ohio give way to the more subtle pancaked landscape of Indiana and western Kentucky. I relished the calm introspection that vast stretches of newly sprouted corn and soybeans could induce. Max, my silver and white German shepherd, kept me company on this journey. I had adopted

Max as a 12-week-old puppy a year before, from a family who had too many dogs and too little time, and he became my constant companion.

Mile markers for Louisville began to pop up as soon as I crossed the Indiana state line. Soon after, I noticed the green tourism sign just short of the city limits for the Louisville Zoo. For the remaining thirty miles, the jumbled pieces in my brain began to shuffle together into an idea, the image of the zoo sign flashing through my head. *Louisville Zoo. What would it be like to work in a zoo?* Visions of elephants, tigers, and chimps emerged as the allure of such a job loomed before me, a salve to the uncertainty commandeering my thoughts. I loved animals, and wouldn't the rest sort itself out?

I had always been drawn to zoos, visiting the Bronx Zoo regularly on trips into New York City from my childhood home an hour away. The animals captivated me, even though I grieved the plight of the big cats as they paced their cement cages, still in use in the 1950s. During my study abroad program in Europe, my first stop in every city I visited was always the zoo. While other students headed to museums for art immersion, I compared the lion exhibits of Paris, London, Barcelona, and Brussels.

By the time I reached the farm and my friends, a plan had begun to emerge. *Could I get a job as a zookeeper?* Maybe then I could see if working with animals was really something I wanted to do. *But what were my qualifications for such a job?* I had no degree that related to animals. *How would I convince anyone to hire me?* But I knew I had to do just that.

---

Upon calling the zoo the day after my arrival at the farm, I was informed I had to talk to the head curator to request a zookeeper application. The next morning, as the sun inched above the cornfields to the east, I retraced my steps north on 42.

The zoo was small but clean and housed collections of most of the animals I expected. The layout consisted of areas divided by habitat and geography. Most of the enclosures were empty at this early hour, the animals probably still eating their morning meal inside.

I explained my mission to the receptionist in the front office and asked for a job application. She pointed to a closed door and the sign "Head Curator." "Go ahead in," she said. "Mike just got back, and he's the one to talk to."

"Come in." A man's voice answered my knock. "What can I do for you?" A six-foot Grizzly Adams doppelganger stood up when I entered. Wearing a khaki shirt with the zoo logo on the pocket and matching pants, one belt loop sagging under the weight of a metal ring laden with a trove of keys, he looked every bit an animal curator.

I introduced myself and stuck out my hand. "I was wondering if you had any job openings, you know . . . like keeper positions?"

"Have a seat. Maybe you can tell me a bit about yourself." This was easier than I had thought. Mike settled back down in his chair, his bulky but solid build speaking of manual labor mixed with a healthy appetite. "Why do you want to work in a zoo?"

Before I could answer, he popped out of his chair to pick up a tied burlap bag from the floor nearby. "Excuse me! Sorry, but I left an animal in here and should transfer him to a crate." He handed me the bag. "Can you hold onto him while I go find it?" I was confused but did as I was told, taking the bag from him. "Tell you what. You can hold him until I come back. He'll probably wrap you, but don't worry. He's not poisonous—a boa constrictor—won't even bite. Probably someone's pet who wandered into this lady's backyard." With that, he opened the sac, revealing a curled-up nautilus, its head invisible amid the chaos of its brown and black pattern. He reached his meaty hand down to grab the snake's head and ease it out into the open. "Here, wanna hold him?"

*He wanted me to hold a snake?* My muscles tensed as my body prepared its fight or flight response. *When should I tell him I'm terrified of snakes?* The zoo had no snakes. I'd made sure of that.

Lifting it gently, he moved it toward me. The tongue was in perpetual motion, in and out. I slowly placed my fingers behind its diamond-shaped head. The animal rose upwards, coiling itself around my arm. I was paralyzed. "Nice and snug," said Mike. "Don't squeeze too hard." I wondered if he was talking to me or the snake. "It's a young 'un, not quite full grown. Poor fella must be scared. We'll find a good home for him now."

As he was talking, I began to relax, becoming used to my new arm bracelet. Fear was giving way to something close to awe. *Not full grown, a young one.* I sensed the snake's fear, so close to my own, clinging to me as if I were its lifeline. The idea that it was young gave me courage. With my left hand, I stroked its back with my fingertips, expecting a slimy skin but amazed at how soft it was.

Returning with the crate, Mike unwrapped it from my arm and folded it inside. The snake curled itself into a tight coil and watched us, tongue never ceasing its flicks.

"What will happen to him?" I asked, the boa's future becoming my concern.

"I'll contact zoos with reptile exhibits and ship it off as soon as we can." Mike stared at me for a minute, then, remembering why I was there, said "You seem to like snakes. Too bad we don't have them here, but we have plenty of other critters you'll enjoy. I think I'll start you with the elephants—tomorrow if you're up for it."

I looked at the boa looking back at me, his soulful eyes seeking a connection on a level I had just begun to grasp.

---

My assigned section at the zoo was the Elephant Area, so named because that species was the signature, and of course largest, critter in there. But there were other species to take care of as well. Some animals seemed harmless but weren't, like the giraffes with their slow-motion movements but ability to deliver lethal kicks in all directions, and the white rhinos, with myopic vision but with the ability to charge at over 30 mph.

The animals I enjoyed watching the most were the birds, emblems of liberty and self-reliance. Despite the confines of the aviary, they could fly, eat, and breed at will. Although the larger species had to have pinioned or clipped wings to prevent long-distance flight, at least they lived outside, safe from harm. Because most of the keepers had little interest in the birds, they were happy to let me take over their care.

Although I oversaw all the African species, it was the elephants who became my mentors. The two African adult females—Mary and Laura—showed me, despite the role many zoos play in conservation, why some creatures should never be held in captivity. In the case of elephants, their size and intelligence work against them. They know things. Like how to escape from an enclosure, how to best a keeper who might turn his back on them for a second, how to use their trunk to grab a bucket and fling it across a room, how to break through a barricade. Of course, a city zoo with limited space can't allow these behaviors to happen. So, elephants must be shackled, one front and the diagonal back foot encircled with iron anklets, whenever they are inside. This situation still exists in zoos today, especially when the elephants are brought into their enclosures at night. If the zoo has enough outdoor space, as was the case in Louisville, the shackles can come off when they go into their yards.

Both Mary and Laura had been circus elephants, so their wildness had been literally beaten out of them long ago. They were never anything but compliant and obedient with me, so I had only to lay the mandatory bull hook gently against their ears, more as a rudder than a disciplinary tool. Their trunks served as noses, mouths, and hands, and I marveled when they used them to travel over my body, searching for treats or to simply connect. Their trust in me was a sacred gift. Feeling their trunks greet me each morning kept me returning to care for them, even though I was having so much trouble condoning their presence at the zoo. What had happened in their past gave me nightmares when I allowed myself to think about it. Mary and Laura convinced me that animals in the wild had to be saved so they wouldn't end up here.

---

By early spring, the novelty of life as a zookeeper began to wane. As my sympathy for the animals grew, it became harder to accept the way things were. Powerless to change the injustices I saw in their care, I lamented not having more of a decision-making role.

Mike took me under his wing over time. When he tired of my many questions and insatiable curiosity about animal behavior, he introduced me to the zoo library, where I retreated every lunch hour. I perused everything I could find on wildlife, not only the animals we housed at the zoo but also other species that were in decline. During my college years, I had never been this excited about learning. I was finally unearthing a destination, or at least a path that might lead to one.

Although I loved working with mammals, seeing them in captivity and being separated from them by bars, cages, or moats, was hard. The zoo library contained a feast of information on all the resident species as well as wildlife from around the world. Most days I was the only one in the library during lunch break, with Mike occasionally dropping in to recommend a book or browse the shelves for himself.

My fascination with birds grew. The longer I observed and cared for the ones at the zoo, the more I wanted to learn about them. I researched all the resident avian species but was most intrigued by birds of prey. We didn't have any in Louisville except Griffon Vultures from Africa, and I asked Mike why there were no eagles, hawks, owls, or falcons.

"They don't make good exhibit animals—super boring when they're in captivity. They are great to watch when they're wild, soaring, chasing prey in the air, or snatching it on the ground, but put them in a cage and their vitality disappears. It's almost depressing watching them just sit there knowing they can't hunt anymore." I pictured the raptors I had seen flying over the fields where my father and I would ride in early morning. To me, they had been the embodiment of freedom, of power. I could see his point; zoos were no place for that spirit. Whether on the wing or perched regally on a branch overlooking a field, marsh, or river, the falcons, hawks, and eagles demanded liberty without restraint.

I read about the sport of falconry, in which humans trained various species of birds, usually the fastest members of the Order Falconiformes, to be their partners in the hunt. This seemed the epitome of a human bond with an animal—a pet that retained its wildness. Although hawks and eagles were also used in falconry, the birds of choice were falcons, the fastest species of the order. Of these, Peregrines appeared to be the most popular due to their exceptional speed.

While pouring over the bird journals in the library, I ran across a project started two years earlier at Cornell University. Determined to reverse the lethal effects on Peregrine Falcons of DDT—a persistent pesticide widely used since the 1940s until its ban in 1972—an ornithology professor and falconer, Dr. Tom Cade, had initiated a captive breeding and reintroduction program in the Northeast known as The Peregrine Fund. This was everything I wanted to do: help a species recover, do conservation work on a hands-on basis, and observe birds in the wild.

# — Chapter 3 —

# Peregrine Teachers

From my seat in the zoo library, vivid images of falcons, relishing freedom for the first time, inspired me as I read more about the Cornell project and the professor who had created it.

I learned that by the 1970s, DDT had ravaged—along with Bald Eagles, Ospreys, and Brown Pelicans—many Peregrine Falcon subspecies, resulting in the disappearance of the entire eastern population. To restore the falcons to their natural habitat, Tom Cade sought to establish a captive breeding program, using subspecies from other parts of the country as well as birds donated by falconers, for the breeding stock. There was no precedent for such a large-scale captive breeding project for raptors. There was no information on how to keep wild falcons—used to flying free at great speeds, killing prey in mid-air, and living on cliffside ledges known as aeries—in secure flight cages while enticing them to breed and lay eggs. Falconry as a sport involves keeping the falcons fit and ready to hunt, but breeding is a whole different challenge. The intricacies of the project involved much trial and error. Every bird had a different history and set of issues to be considered.

When Cade initiated the program, he had to secure the blessing of Cornell—who had recruited him from Syracuse University in 1967—as well as its nonprofit funding arm, the Laboratory of Ornithology. He needed financial support, wildlife permits, a staff with expertise in falcon husbandry and behavior, and most of all, a place to house the captive breeding venture. The Hawk Barn—a voluminous structure across the street from the Lab—was constructed to be the center of operations. Housed in spacious flight cages holding one or more falcons, some of breeding age and some juveniles waiting their turn, every bird represented promise. Their success in reproducing and surviving hinged on the balance of luck and the ingenuity of the Peregrine team of researchers.

I spent hours trying to imagine the challenges Tom Cade had faced during those early days. He had a vision, but putting the pieces together to make it real was the critical part.

To launch his program, Cade needed someone to manage the daily operations of the enterprise who understood birds of prey, was familiar with captive breeding and falconry techniques, and shared his passion for the cause. Jim Weaver arrived at Cornell to fill this niche as Cade's right-hand man. The two men, ostensibly different in every way, had a mutual admiration for one another based on firmly established territories and complementary skills. Phyllis Dague was brought on to the team to oversee the administrative aspects of The Peregrine Fund, and along with Willard Heck, a graduate of Cornell's Natural Resources Department, assisted in the captive breeding program in the Hawk Barn. While others came and went over the years, these four were the heart and soul of the reintroduction program that saved the Peregrine.

By the time I arrived at Cornell in 1974, the Peregrine Falcon releases had been operational for several years and were in their mature stage, the researchers having made and rectified many of the mistakes typical of a startup program. Some of these involved the usual issues surrounding captive breeding, like the intricacies of artificial insemination, the delicate balance of incubating the eggs at just the right temperature, and the imprinting of the falcons on humans. And some involved predicting how young falcons could survive in the wild without their parents around to protect them.

Because wild falcons usually nest on the sides of cliffs, it seemed obvious that the captive-bred young should be released from sites that resembled their native habitats. One of the hard lessons the researchers learned early on was that predators come in all shapes and sizes. While the cliff sites limited predation from raccoons and other mammals, nobody had predicted that Great Horned Owls would wreak havoc on the program. A large avian predator of fierce reputation, the owls found the young parentless falcons to be easy prey. The obvious solution was to trap and remove the enemy birds from the areas where the chicks were being killed, but within a day or two another pair of owls would come in and continue the pillage.

In the end, it was best to relocate the release sites from the ravines and cliffs of northern New England to cities, where bridges and skyscrapers made much safer and more effective homes. Pigeons became the preferred and plentiful food source for the young fliers, easy pickings for those who were just learning their skills. It was this procedural change that would not only save the program but also allow falcon sightings to be enjoyed by city dwellers for years to come.

---

On a clear September morning in the 1980s, a pedestrian on the Manhattan shoreline of the Hudson might have been treated to a mesmerizing sight—one not included in the *Big Apple* guidebook—of a grey torpedo-like bird hurtling itself from the abutments of the George Washington Bridge onto a slow-flying pigeon. The attacker would then swoop upwards with prey in its talons and return to its perch hidden below the lanes of cars crawling between New Jersey and New York. Even if they didn't recognize the killer bird, the person would have known they had seen something unusual, and stupefying.

Urban releases didn't start until 1980 when the first pair of falcon nestlings were placed on the Manhattan Life building near West 57th St. Later, young Peregrines were released from many of the bridges crossing the East and Hudson Rivers as well as from other skyscrapers in the city. On any given day, one could look up between the towering buildings flanking the streets of Midtown to glimpse a soaring falcon searching for its next dinner.

---

The more homework I did on the history of The Peregrine Fund—its mission, its setbacks, its successes—the more courage I found to follow my own dreams. By May, I decided to leave the zoo when my annual contract was up in the fall, apply to Cornell, and do graduate work on Peregrines with Dr. Cade. I was soon to discover that wanting something and attaining it were two different things.

## — Chapter 4 —

# Climbing the Hill

Luck and a gentle boa constrictor had secured me the Louisville Zoo job. Without a science degree and the courses that accompany it, being accepted into a Cornell graduate program would not be as easy. That spring, while still working at the zoo, I sent off my application without much hope of a positive reaction. I was surprised. The return letter wasn't a definite no but rather an acceptance with a condition: my entry to grad school would be contingent on completing all the requirements for a second undergraduate degree, this one in biology.

As if that weren't enough of an obstacle, the letter also informed me that once I had completed the necessary coursework, I would still need to find a professor willing to mentor me and serve as my thesis advisor. My research on Tom Cade and his Peregrine project had convinced me that I wanted to study with him, but *he* didn't know that yet. And how could I hope to convince him if I didn't meet even the minimum requirements? I knew I needed to prove myself in science before he could even consider taking me on as a graduate student.

The road ahead seemed formidable and infinite, and frankly, scared me to death. The college catalog listed the classes I would need to take, many of which I had already stumbled over, and then abandoned, in my previous pre-med effort several years earlier. How could I reenter the physics classroom or wrestle with organic chemistry when the names alone made me quake?

Reexamining my decision to tackle science again had become a daily mind game during my final days at the zoo. Seeing the elephants, birds, and my other charges confined to their captive worlds every day convinced me I couldn't turn my back on the challenges in wildlife biology and the needs of the natural world just because I was afraid of how hard it was going to be.

In July of 1974, a year after starting at the zoo, I said goodbye to Mike, other close colleagues, and my favorite animals, like Mary and Laura. With Max and Moses—a black-and-tan collie mix I had rescued after someone dropped him off at the zoo a few months earlier—I headed east towards Ithaca, New York, in the southern tier of the state, an area unfamiliar to me, toward a university I had never seen.

As I approached Cornell, brimming with enthusiasm and little else, I knew that life was about to change if I could just read the proper tea leaves. I had been accepted as a "special student," a euphemism for a nondegree program. This meant I could take the courses required for the undergraduate biology major but would not be part of any graduate department. This was akin to being on the outside of a candy store looking in, with nose pressed to the glass, hoping for admittance to sample the treats inside. There was no funding, no housing, no meals provided, no advisor, no student community, and no set curriculum. The sense of not belonging added to my feelings of isolation. More often, however, it served as my motivation to stay the course. I had to trust the candy was within reach.

Luckily, I was able to locate an affordable room in a house with five other students. My savings from the zoo job, combined with a small legacy from my grandmother, helped me with food, rent, and books, and would be just enough to cover two semesters of courses. If I hadn't been accepted as a full-time graduate student after that, I knew I would have to face reality and choose another route.

---

Cornell made a dramatic first impression on me with its unique topography and physical command of the area around it. Two fast-flowing streams, which cut deep gorges accented by waterfalls, formed the border of a campus that resembled a hilltop kingdom. The university's layout looked like the front of a multi-tiered cake, starting from Cayuga Lake as the bottom layer, and rising one story at a time until reaching the top tier of the main campus.

My apartment was near the bottom of the cake, in the downtown area, so I had a long climb to the academic quads where my classes were held. I had brought my two dogs with me from Kentucky to Ithaca, so the fact that my landlady and roommates were animal lovers was a godsend. Max and Moses adapted easily to university life. Cornell had a rule that dogs could roam unleashed on campus. Rumor had it that some benefactor had made a large donation to the university on the condition that a leash law never be imposed. I didn't know if that were true and didn't dare ask. The dogs had to come with me to class, since I didn't have time to return home to walk them during the day. It was a relief to see many dogs playing or sleeping outside on the quads while their owners attended classes. Because I never had more than one class in the same building, I was inside for less than an hour at a time. This allowed me to check on them often between periods, give them water, and ensure they were behaving themselves. Friends never had a problem finding me since the pair wouldn't be far from the building they saw me enter.

People might not have known me, but they knew Max and Moses. Most of the time, everyone was careful not to let them into the buildings, as "No Dogs Inside" was the only rule the college enforced. But it wasn't a perfect system. Max's loyalty to me was built into his German shepherd DNA, his mission in life to be by my side whenever possible, a trait I came to depend on. I endured several embarrassing moments when he and Moses, the constant sidekick, would find their way into a building, locate my class, and come for a visit, much to the delight of the other students and sometimes of the professor.

Because it was going to take at least an extra year to fulfill my requirements, I signed up for every science course I could fit into my schedule to speed up the process. At one point I was taking physics, organic chemistry, genetics, physiology, and ornithology simultaneously, which precluded sleep, meals, and any life outside a classroom or lab. I existed in a perpetual state of panic as tests, lab reports, and homework threatened to bury me. Looking back on that time still engenders panic in my gut. There were many days when I was tempted to quit, to surrender to the science gods who didn't want me on their team. A combination of my friends, who stepped in and talked me down from my cliff, and my pride, which couldn't admit to another failure, kept me going.

Even though ornithology wasn't required, I included it to provide a bright spot in my crazy life and to help me survive the other classes. When my confusion about acceleration and velocity in physics and structures of organic compounds in chemistry reached its peak, I could count on taking an afternoon bird walk with my ornithology class. I immersed myself in the woods of Ithaca where the hardest task, reminiscent of my introduction to birds in Ohio, was identifying the various species of warblers passing through in migration. I quickly mastered the common ones like Yellow Warblers and Common Yellowthroats, and eventually learned to recognize the colorful Northern Parula, the more subtle plumage of the Yellow-rumped, and my favorite of all, the striking Blackburnian, among others. When it came to bird study, whether outdoors or in the classroom, I was a sponge, eager to absorb every detail thrown at us about their anatomy, behavior, and even their complex taxonomy. These afternoons, surrounded by these animals in their habitats, convinced me that this was where I belonged.

Just because I had reentered the world of science and was surviving didn't mean the material had become any easier for me. No new science genes had arisen in my body since freshman year, and the more the material delved into the molecular level of cells and their chemical and physical processes, the less confident I felt. What I did learn to do was work hard. It seemed to take me longer than my classmates to master the concepts. Most of them had chosen science for the love of it or because they were good at it. I, on the other hand, needed to focus on the road ahead or risk becoming stuck in a quagmire of equations, formulas, and experimental data. How does this relate to working with animals, I would ask myself, and when I couldn't answer, I resorted to willpower. The images of Mary and Laura, with their intelligent elephant eyes gazing down at me as I shackled their legs for the night, were certainly contributors to my doggedness, urging me forward.

It was impossible to distinguish between all the schools and departments at Cornell without advice, and since I had no advisor, I relied on others who were enrolled in the programs. I met several graduate students in the Department of Natural Resources, which focused on environmental issues and management. With its applied approach to wildlife science and my desire to work directly with animals, it was a better fit than the more theoretical Division of Biological Sciences.

The friendships formed in my ornithology class led to a chance encounter with someone who had a lasting influence on my life beyond Cornell. One night, a group of us gathered at a bar where I met Larry Morris, a grad student in Natural Resources. Larry had arrived at Cornell two years before me and was working on his PhD in natural resources conservation, one of the specialties in the department. Through him and our burgeoning relationship, I discovered a world outside the textbook and classroom, and life at Cornell took on a shinier gleam.

# — Chapter 5 —
# Opportunity Knocks

Although working with Tom Cade and his team at Cornell was my goal, I had no idea how hard it could be to make the tryouts. Convincing Cade to take me on as a grad student, even once I completed the prerequisites for a biology degree program, was a pipedream. Self-promotion is not my strong suit, but I knew I shouldn't be at Cornell if I was unwilling to try to make things happen. It was time to introduce myself to the man I had come to work with. After setting up an appointment, I drove out to his office at the Laboratory of Ornithology. I had nearly completed my one-year stint as a special student, a solid transcript of science courses within my grasp.

Dr. Cade's office was tucked into a wing of the Lab building, separating him physically from the bustle of the facility. His geographic position reinforced what everyone knew: The Peregrine Fund was a world unto itself and seen as a separate entity managed by Cade and his team. I introduced myself to one of the primary members of this team, Phyllis Dague, his administrative assistant, whose desk sat outside his open door. Her warm greeting helped ease my nervousness and gave me the reassurance I needed to enter his office.

A trim athletic-looking man, Dr. Cade was sitting in his chair behind the desk. He seemed out of place there, but perhaps it's because my mind flashed instantly to the photographs I had seen of him standing proudly with a falcon on his wrist. When I walked in, he leaned back, clasped his hands behind his balding head, and fixed me with a stern look. When he began to speak, his Texas accent, muted by years in California and the Northeast, but still unmistakable, offset my first impression that this man was scary and unapproachable.

My introductory meeting with him was cordial but disheartening. He was a forthright man, soft-spoken and kind, who didn't bother with small talk. When I explained that I had come to Ithaca in the hopes of working with him on a raptor study, he replied politely that he was flattered but couldn't take on a student who didn't have a funded project.

"Could I help with the Peregrine project? I'm willing to do anything, be a gopher, assist with releases, anything else?" I countered, trying to sound enthusiastic and helpful.

"I'm afraid that program has all the students we need. There really isn't room."

"Maybe I could work with Ospreys?" I refused to give up even though I was embarrassed as the tone of my voice edged toward desperate.

"I already have a graduate student working with them. He's doing an egg relocation project in Connecticut and Long Island, so that area is covered right now."

His expression displayed sympathy and regret; it was apparent he took no pleasure in rejecting me. When he looked over my transcript with the long list of courses I had taken, his eyebrows rose toward his smooth forehead. His tone was gentle as he said, "If anything comes up that has some funding attached to it, I'll be sure to let you know." I left his office deflated, aware that all the work I had been doing might not be enough.

When I first arrived in Ithaca, my plan had been to ride on Dr. Cade's coattails while I learned from him. I now understood I needed a specific strategy to make this happen, and it would be up to me to design one. I couldn't allow myself to accept defeat after spending so much effort to reach this point. As it happened, my timing had been fortuitous.

Feeling particularly low and in need of distraction in the following weeks, I poured out my frustrations to Larry, who by this time had earned the dubious honor of being my main support system. He was always successful in turning around my defeatism by encouraging me not to abandon my ambitions, despite all the setbacks. That weekend, he happened to be hosting a former college friend, Amos Eno, who was visiting from Washington, DC. Amos was taking a one-year sabbatical from his job with the US Department of the Interior to pursue a master's degree at Cornell in natural resources and public policy. Larry invited me to a get-together at his apartment where several people from Natural Resources were coming over to meet Amos. Even though I wasn't yet a member of the department, he introduced us.

During our conversation, Amos asked me what I was studying. I told him of my fruitless efforts to work with Tom Cade and my need to find a project and funding before he could accept me. After months of wading through science facts and figures, questioning my abilities all the while, my life was transformed by Amos' next words: "Would you ever want to work with Bald Eagles?"

"Eagles! Are you kidding? Sure!" I answered, having no idea where this was going. The possibility of working with eagles lay well outside the confines of what I considered my reality. The majestic dark bird with its singular white head and tail and infinite wingspan represented the pinnacle of the avian kingdom, connoting power, mystery, and mastery of the skies—a species far too sublime to imagine. My mind might have traveled, at this point, toward visions of myself studying Bald Eagles, saving a species in decline, making a difference. But I never considered our conversation at the party anything but idle chitchat and didn't allow myself to move further down the road of possibility that evening.

But my fortunes were beginning to turn. In 1975, with the nation's bicentennial in sight a year later, rumblings among New York state and federal wildlife agencies centered on doing something to forestall the demise of the Bald Eagle. Unbeknownst to me, Amos' boss in Washington, Nathaniel Reed, was the Assistant Secretary of the Department of the Interior for Fish, Wildlife, and Parks. A proposal to reintroduce Bald Eagles into New York State had recently come across Reed's desk, sent by Ogden Reid, Commissioner of the Department of Environmental Conservation (DEC) in Albany. New York State wanted to use the same techniques as those Tom Cade had developed to release Peregrines. The Fish and Wildlife Service, overseen by Reed, had been reluctant to approve the project since the state didn't have anyone on staff with raptor experience. Consequently, the proposal had been denied and set aside. Projects involving species such as eagles, federally listed as endangered, could not be undertaken by a state agency without Interior's approval.

Until that night, there hadn't been a way to counter this stumbling block. Once Amos realized I needed a funded project to work with birds of prey under Cade, he began to put the pieces together. A graduate student willing to take on the daily operations of the Bald Eagle reintroduction, while Cade continued his work with the Peregrine releases, might be the solution needed to protect the national bird before the bicentennial. Amos contacted Nat Reed and ran his idea by him. It only took one phone call to Tom Cade from Amos and Reed, outlining the Bald Eagle program, to turn the tide in my favor. The Bald Eagle proposal was back on the table.

I heard from Cade the night after he received the request from Nat Reed. One of my housemates yelled upstairs to me that I had a phone call from a Dr. Cade. I knew that Amos had been working behind the scenes, but I didn't know that Cade had already been contacted. In his typical laconic tone, he said he had a proposal for me and invited me to his office the next day. I didn't sleep that night.

When I entered, Cade was on a call and motioned for me to sit down. I did but not easily. I felt like I was ready to explode. Finally, he hung up and smiled at me. "I have some news that I think you might like. There is a proposal that has been sent my way from Fish and Wildlife. I believe you know Amos Eno? He has been helping us secure federal funding for the Peregrine program for the last couple of years. Now Interior wants us to consider reintroducing Bald Eagles in time for the 1976 bicentennial. We need a graduate student to do the field work, and Amos and I thought of you."

I wanted to leap to my feet and shout but knew one didn't act like that with Tom Cade. Nobody had more passion for saving birds of prey than he did. He had created the Peregrine project, had convinced Cornell to provide a home for his program, had climbed the academic ladder to a full professorship, and through determination had made his mark on the bird world. But dancing around his office or screaming my joy from the rooftop were not going to earn his respect. Instead, I calmly thanked him for the opportunity to work with him, told him how much

I admired eagles and how thrilled I would be to be part of this project. As I struggled to keep my emotions under control, we discussed the logistics of the program before us.

Meanwhile, my brain was sizzling. *How did I find myself in this position?* Up to now, my dreams had been merely dreams, only that, certainly not reality. Once Cade had been assured of my willingness to sign on to the eagle work—he must have known looking at my face that it was hardly a tough sell—he agreed to supervise the program on behalf of Cornell. As I stood up to leave, he also rose and held out his hand, his smile creasing his chiseled features. "I'm glad this has worked out for you, Tina. You've worked hard and it's paid off."

---

It wasn't long before the political complexities of the program became evident. Because New York's DEC had first proposed the reintroduction, the state would be a partner in the project. Cornell would initiate the field work and manage the project for the first couple of years, US Fish and Wildlife and DEC would provide financial support, and I would be the research biologist working with the eagles under the watchful eye of the Peregrine team. Accepted by Tom Cade into his cadre of graduate students, I finally had an opportunity to do applied work in bird conservation. Now I only had to prove I was qualified to be here, not only to Cade and his team, but also to myself.

The project seemed vague and risky, but well-funded. Everyone wanted something to happen to bring back the eagle before July 4, 1976. For some, focusing on the date was a way to draw attention to conservation agencies and their programs; for others, restoring the eagle was symbolic of a nation seeking to right a wrong by avowing its commitment to the environment and to the bird who epitomized freedom and resilience.

Time was tight, though. If it had gone out to bid for someone to take this on, I wouldn't have stood a chance. I had never worked with eagles, was just starting my graduate studies, and my only experience caring for wild animals was a year spent in the Louisville Zoo, where I had been the elephant keeper! But the state and federal officials were in a hurry, and I was there, ready and willing, and none of us really read the fine print about this project.

My head was so consumed by the fantasy of saving the eagle, I neglected to think through what this would entail. I had no conception of what lay ahead, what challenges I would face, and what the consequences could be if things didn't work out. Seduced by its possibilities, I was diving into this high-profile program without checking the depth of the water.

In six months, I would find myself on a hilltop having sole custody of two Bald Eaglets whose future survival lay in my hands. It had been a journey to reach this point, and the pressure to ensure the birds' survival was paramount. All the missteps and faulty decisions I had made along the way fell by the wayside, no longer important or relevant. I was here now and had been given an opportunity to restore the Bald Eagle to its rightful place in our northeastern skies.

## — Chapter 6 —

# Project Eve

I had arrived at Cornell in the right decade. The 1970s were a time of awakening when environmental threats to water, air, and wildlife became rallying cries across college campuses and in the halls of government. Responding to the public's concern, in 1970, President Richard Nixon established the Environmental Protection Agency to monitor pollution, a move designed to keep the country safer for wildlife as well as people. The Endangered Species Act followed in 1973, making the Bald Eagle one of the first animals protected by law. Nixon may have encountered many issues during his presidency, but his environmental record was impressive.

Bald Eagles, revered by indigenous people, had enjoyed a peaceful coexistence with humans until the nineteenth century when they became targeted as morally depraved scavengers and predators of livestock. People resented them as competitors for fish, a dietary staple of both human and eagle, especially in Alaska where the birds were so plentiful. By the early 1900s, the species was persecuted into near extinction in many areas by shooting, destruction of nest trees, or habitat decimation, ultimately resulting in passage of the Bald Eagle Protection Act of 1940.

Eagles were still not safe from human ignorance. During this period, DDT was introduced to protect the military in World War II. This organochlorine chemical pesticide, whose long-lasting effects were heralded as a solution for malaria-carrying mosquitoes as well as those that were a nuisance, presented a second threat to the Bald Eagle. Its population numbers, which had only begun to recover, decreased precipitously before anyone understood the cause.

By 1963, the Bald Eagle in the lower forty-eight states had declined from about 400,000 birds in the 1800s to only 417 nesting pairs. Without warning, DDT had begun to wipe out Bald Eagles, along with many other avian predators atop the food chain. The eastern regions of the United States witnessed catastrophic declines in their populations due to the heavy application of DDT there, while the midwestern and Alaskan birds were less affected.

As early as 1962, biologist Rachel Carson sounded the alarm about the long-term effects of the pesticide on birds in her famous book *Silent Spring*. Scientists at Patuxent Wildlife Research Center, the research facility of the USGS, had confirmed what Carson suspected: DDT disrupted calcium metabolism, causing the thinning of eggshells in birds like eagles. While the contaminated birds could still breed and lay eggs, no young could hatch due to the cracking of the shells when the parents sat on them during incubation. By 1972, a decade after Carson's landmark book, DDT was banned, but the damage had been done. How could the nation celebrate its two-hundredth birthday without honoring and, more importantly, saving, its national symbol? The impending anniversary of the country had galvanized public officials to take up the eagle's cause once more, just as people had in the 1930s.

By 1975, news of the Peregrine Falcon program's success had spread far beyond the Cornell campus. Even before the Peregrine releases were moved to cities, people were aware that this was a groundbreaking project that had changed the course of an endangered species' future. A falconry technique known as "hacking," which involved removing young birds from their parents, caring for them until they were ready to fly, and releasing them back into the wild in another locale, was the mainstay of the project. Since many of the release sites were within New York, the state's Department of Environmental Conservation (DEC) was familiar with the success of the program and the publicity it had engendered.

Four years after the first Peregrine releases, attention turned toward the nation's bicentennial a year later. The need for a national effort to conserve wildlife, and particularly our national bird, soon became clear. New York's DEC was initiating an endangered species program that year, and there would be no better way to kick off its inception than a high-profile effort to bring back the Bald Eagle, just as Cornell had done with the Peregrine. Under the terms of the Endangered Species Act, the Fish and Wildlife Service had to be involved in such a program and approve any actions to work with Bald Eagles. Knowing that Cornell was the obvious home for such an undertaking, Fish and Wildlife and DEC asked Tom Cade to consider conducting a Bald Eagle hacking program.

Although the Bald Eagle project would involve a good deal of guesswork, it started off being based on what had worked with the Peregrine releases. Issues would arise frequently, however, as everyone recognized that what worked for a falcon often wasn't naturally transferrable to an eagle. The two species were different in almost every way: size, development, prey selection, and hunting behavior being the most obvious.

Because there was no roadmap, the eagle program counted on the Peregrines to lead the way. The falcons had already demonstrated that predation on vulnerable young birds, no matter that they would become predators themselves one day, is a force to consider. Without parents to protect them, even young eagles, despite their larger size, could be threatened by predators like owls and raccoons when in their nest, or coyotes and foxes if they were stranded on the ground.

The national spotlight shining on this project placed pressure on Tom Cade and his team to ensure its success. Despite the experience gleaned from the Peregrines, there were many unknowns and invisible potholes to avoid. The summer of 1976 would be a dry run, a chance to figure out if hacking would work with eagles, and a chance to resolve any problems before expanding the program's scope to reintroduce more birds.

---

Often, bird reintroductions rely on taking eggs or nestlings from breeding adults and placing them into the nests of nonbreeding birds who adopt them as their own. This technique was used by Paul Spitzer, one of Cade's students, who transferred Osprey eggs from healthy adults in Maryland to nonbreeding DDT-contaminated birds in Connecticut and the eastern end of Long Island. The problem with both Peregrines and

Bald Eagles, unlike Ospreys, was that there was no adult population left in the Northeast to act as foster parents. Another method was needed.

Although the hacking procedure, effective with the falcons, was deemed the optimal way to reintroduce Bald Eagles, the Peregrine project had one main advantage. The world of falconry is relatively small and close-knit, and most falcon owners wanted to see Peregrines brought back to their former habitats once DDT was banned. Tom Cade was able to use birds donated by other falconers as breeding stock. He then launched a captive breeding program and hacked the young progeny of these donated birds.

By contrast, there was no captive breeding program for Bald Eagles in 1975. Although some falconers kept Golden Eagles for hunting, few had Bald Eagles. Because their population was so low in the lower forty-eight states, it was difficult to obtain a permit to keep them. They were also expensive to maintain due to food and space requirements, and they didn't make for ideal hunting companions. Young eagles for hacking had to be found in the wild. Alaska would have been the likeliest source of Bald Eagles because of the large number of nests there, but there was concern that this different subspecies might result in mixed gene pools. Therefore, it was decided early on—the result of many meetings between the Cornell and Fish and Wildlife biologists—that the eaglets to be released in New York State would be transferred from nests in the Great Lakes states where DDT hadn't taken a toll on their populations. Wisconsin, Michigan, and Minnesota were the likely candidates.

The political dance required to transfer an endangered species, much less the national symbol, from one area of the country to another, was a choreographer's nightmare. After much discussion through the winter of 1976, the US Fish and Wildlife Service agreed we could take chicks from eagle nests in Wisconsin where the population was healthy. Of course, this had to be negotiated in a sensitive manner, so it wouldn't appear that we were robbing the Midwest of its birds. But it was known that healthy Bald Eagles, such as those around the Great Lakes, often lay three eggs but can raise only two chicks due to food supply limitations. It made sense to "adopt" these third chicks and ship them to New York where they would be raised by a foster mother—aka me.

Given that nobody knew whether young eagles could survive without their parents' supervision, it seemed wise to start with just two birds in 1976. This first year was to be an experiment designed to test the hacking techniques and iron out any problems before more birds were transferred into New York.

---

The selection of a New York release site was the next dilemma. It was critical that the eagles be kept out of the public eye until they could become independent flyers, so a protected area was required. There had to be a plentiful supply of food easily accessible to young birds, as well as a combination of trees and water. Ideally, everyone wanted assurance that the site would be attractive enough to entice the eagles to nest there later. The species was known to return to their natal area once they were old enough to breed.

Montezuma National Wildlife Refuge, a 10,000-acre expanse of swamp, marsh, and streams located at the northern end of Cayuga Lake near Seneca Falls, fit the bill. It was a refuge managed for migratory waterfowl with plenty of pools available whose water level could be controlled by dikes and seasonal drawdowns. It was forty miles from Cornell, allowing the Peregrine team easy access in the early days of the project. And it was also a historic nesting site for Bald Eagles, active until 1959. Even though none of us knew whether the eagles hacked in 1976 could survive without their parents, much less return to their nest site, it still seemed we should allow optimism to lead the way when making decisions about the project's location.

# Part Two

# 1976

## — Chapter 7 —

# Beginnings

Although I had worked with birds in the past—injured nestlings as a child or the African bird collection as a zookeeper—I had no experience with birds of prey. These predators, with powerful talons and beaks specialized to catch and kill prey whether in midair, in the water, or on the ground, posed a different challenge altogether.

The first order of business before I assumed custody of the eagles, therefore, was to train *me*. I was to shadow the Peregrine crew and learn everything I could about raptors and their care. I knew if I flunked out of raptor boot camp, there would be no eagle project or graduate program in my future; the heat was on.

Cade knew I was nervous about taking on such a big challenge, as exciting as it might be. My feelings were ricocheting in all directions, and I needed convincing. As a first step to launching me into my new world, he escorted me over to the Hawk Barn so I could meet his staff. In retrospect, I'm not sure if he had actually informed them that he was adding to their workload by handing me and my nascent eagle project over to them.

Upon entering the facility, I knew something extraordinary was occurring within these walls. In the ensuing weeks, the Hawk Barn would become my Valhalla. It was there that I discovered a refuge from my academic world of labs and textbooks. The vociferous *scree* of the falcons before mealtime, the odor of their raw chicken and quail dinners, their brownish-black mask and whiskers in contrast with their heavily streaked underparts, provided the empowerment that I was seeking. My reservations did not disappear but instead became relegated to the backseat of my mind behind an eagerness for what was to come.

My gender might not have been a concern for Jim and the rest of the Peregrine team, but this project would be subjected to much public scrutiny. Knowing that women were still considered a rarity in the raptor world, I wanted to be accepted on my merits, not my uniqueness. Tom Cade had only one or two female graduate students, none of whom studied birds of prey.

From the start, I was in awe of the crew who worked with the falcons. Because the Peregrine releases had started five years earlier, the team was ensconced in their program and comfortable with their work and each other. All were expert falconers, most hailing from the western states with years of experience dealing with these birds. The men, especially Jim and Willard, were patient and willing to teach me the ropes of how to handle raptors—respecting their talons and beaks all the while—and what techniques were essential for captive breeding and releasing young falcons. It was up to me to figure out how to transpose these skills onto the raising of Bald Eagles.

Each time I left my apartment and headed toward the Lab of Ornithology and the Hawk Barn, which sat several miles from Cornell's main campus, I felt I was embarking on a daring adventure. As the landscape became more rural and open, excitement careened through my body. Filled with uncertainty and dread one minute, while brimming with anticipation and elation the next, I knew my life was about to change. Even the street address of the Lab—Sapsucker Woods Road—told me I was entering a world of birds and wildness. If I was under the illusion I was crossing the Rubicon in my long quest to work with raptors, I soon was to discover my challenges hadn't even begun. If I had assumed being accepted by Tom Cade as a graduate student meant he was going to lead me by the hand through the initial stages of setting up the eagle program, I was sorely mistaken. It became apparent early on that nobody was offering me training wheels.

---

Although Cade was the academic and professional face of The Peregrine Fund, he was a professor with a full course load and graduate students to oversee. What I didn't understand at first was that he had accepted the request from Fish and Wildlife and DEC to conduct the eagle program while knowing he had little time to invest in it. If I hadn't been so willing and available, so pitiful in my desperation to work with Bald Eagles, he probably would have turned it down. His laser focus on The Peregrine Fund, his own creation and passion, made everything else pale by comparison.

Even with the Peregrine program, much of his time had to be spent fundraising, writing about the project's successes, and dealing with university and government administrators. Although he conferred daily with his staff, the day-to-day care of the birds and set up and monitoring of the release sites fell to Jim, Willard, and Phyllis.

An experienced falconer, Jim possessed a singular knowledge of raptors and the wild nature of birds. A self-taught naturalist from Illinois, he never seemed to let the Ivy League aura of Cornell get to him. Saving Peregrines was what mattered, and he knew more about it than anyone. He became indispensable to Cade, for his instincts and knowledge of falcons proved essential to the success of the program. Since nobody knew the techniques of raising birds of prey better than he did, it was inevitable it would fall to Jim to design and supervise the Bald Eagle project.

Jim was not part of the academic world of Cornell. His connection with grad students was through the birds they were studying. Unless they were working with Tom Cade, they wouldn't have known who Jim Weaver even was, which seems incredible considering how essential he became to the future of Peregrines and Bald Eagles. While Cade served as his students' academic mentor, advising us on classes, thesis topics, and research techniques, Jim guided us on bird behavior and care, on the instincts needed to work with birds of prey.

Luckily for me, he was a born teacher, whose patience and expertise provided me with the on-the-job training I would need to survive the next two years. A reticent man whose spare words exuded calm in every situation, Jim kept me sane. Even with Cade, he was the voice of reason when the older man became caught up in his enthusiasm for his various projects. Physically, Jim reminded me of the quintessential cowboy—rugged, reserved, at one with the outdoor world—belying his midwestern roots. His unruly chestnut mane of hair and bushy mustache, combined with his unchanging uniform of denim or flannel shirts, blue jeans, and work boots, spoke to a lifetime of outside work and climbs to eagle nests and falcon aeries.

I knew long before the eagles arrived that I would be reliant on the advice of those more experienced in raptor behavior. I was supposedly the researcher in charge, but that was a sham since I brought little to the table in the beginning except enthusiasm and a healthy fear of failure. Listening to others who knew more was easy; making mistakes and learning from them was a more difficult but essential part of the game.

## — Chapter 8 —

# Never Look Down

My inexperience kept me from being included in many of the early decisions about where the young eagles would come from and where they would be released. These were left to Tom Cade, DEC, and Fish and Wildlife to sort out. Feeling like I was wasting away on the sidelines while waiting for the action to begin, I satisfied my curiosity to see my future research site by making an early morning reconnaissance trip to Montezuma in early spring.

Even though the refuge stretched out beyond the northern end of Cayuga Lake, only an hour's drive from Cornell, I had never been there. Entering through the main entrance, I was impressed by the endless expanse of marshland and open water extending beyond the dike roads. Dead trees stood like sentries in the middle of the pools, inviting raptors, herons, and other birds to avail themselves of convenient perches from which to survey the water below. Only one building existed inside the gate, the refuge station, but it was empty at this early hour of the day. I had wanted to ask someone to direct me to Clark's Ridge, the hilltop on which the eagle tower would be built. Although I could see it from where I stood on the dike road, a huge body of water lay between us. Without directions, I had no idea how to reach it. The ridge had been chosen as the release site because it was the highest point in the refuge and could be cordoned off from public access. The eagles would have a perfect view of the Storage Pool (later named Tschache Pool), a large impoundment at the base of the hill. They would also be protected from curiosity seekers since nobody would be able to locate the farm roads leading to it. Everyone was aware that all decisions had to be made bearing in mind the privacy and safety of the birds above all else.

As I looked across the pool toward the ridge, I tried to imagine Bald Eagles soaring down from their tower, gliding on their six- to seven-foot wingspan, dipping their feet into the glassy surface of the water, rising with a fish secured in their yellow talons. Then they would fly to a snag to finish the job of ripping through the scales with their giant beaks and, with all weapons working together, tearing at the flesh they had earned for their dinner. Little did I know what lay between this idealized scene and the reality I was soon to face: young eagles, too weak and not feathered enough to fly, learning about their world from the confines of a man-made nest, and from a human eagle mother who knew little of eagle behavior. I sought shelter in imagination in case I lost faith in my abilities to carry off this project. The calm of the refuge on this day was not to be repeated for many months as the burden of responsibility settled on my shoulders and preoccupied my thoughts.

---

The first test of my abilities as a surrogate eagle provider came before arrival day. With no eagle parents around to select a suitable nest tree, we had to build a nest platform where the young eaglets would spend their time.

In early May, after my solo visit to the refuge, construction of the tower began when the New York State Electric and Gas Corporation sunk four telephone poles in the ground on top of Clark's Ridge. Jim went up to Montezuma a few days later and, with the help of some Peregrine crew members, built a wooden platform, just over seven square feet in size, a few feet below the tops of the poles. Using two-by-fours and four-by-fours for support and laying one-by-six planks at right angles on top, spaced about an inch apart to allow for drainage, the platform looked like a raft suspended in midair. The only route to the top was provided by steel pole steps—identical to those electrical workers use—that had been inserted into one of the telephone poles.

By the time I made my way to Clark's Ridge later that same day, the tower was in place, and Jim and the crew had secured the platform to the poles. It was my first time on the hilltop, and I was struck by the views of the vast marsh below, visible through openings in the sparse spring foliage of the trees. But in truth, I was more overwhelmed by the height of the tower before me. Just as my brain was connecting the platform height with the climb up the pole steps required to reach it, Jim turned to me and said, "Let's go up and see where we should build the nest." *Go up?* I realized I had reached this point only by convincing myself that this day would never come. My undoing was imminent.

I never considered I would need to climb this tower every day to observe, feed, and care for the eagles once they arrived. I never thought that my fear of heights could prevent me from ever starting this project. My phobia—admitted to no one—was so severe that I avoided sitting in balconies or getting tickets in the upper grandstand of Yankee Stadium, and never could my brothers convince me to climb a tree. The thirty-six-foot-high tower of telephone poles might as well have been the Empire State Building.

My mind flashed backward to my past fears, all of which had threatened to sabotage me from reaching my goals. The Jersey bull at the farm next door, with its pawing hoofs and steam coming from his nose, who wanted to stop me from entering the farmyard. My horse, leaping over fences as I clung to his mane, hiding my terror lest I be banned from those sunrise rides. The boa constrictor, whose slithery body threatened my dreams of being a zookeeper. The gauntlet had been laid down each time, as it was now. If I showed Jim the dread I harbored inside, I was doomed. Somewhere in the depths of my psyche, I needed to find the courage I wasn't even sure existed. I had found it once, so I had to believe I could find it again.

Jim made his way up the pole like the practiced climber he was while I remained paralyzed below him. When he was almost to the top and about to turn around to witness my stricken expression, I began to climb. My hands were sweaty and my knees weak. Each step called for my tenacity and backbone to materialize and save the day. I had to show him that I was just like the others, had no reservations, could do this job as well as the guys, if not better.

Aside from Phyllis Dague, there were no women involved in the Peregrine program at that time, and few were working with raptors of any kind in the United States in 1976. Although Jim and the others on the Peregrine team never made me feel being a female was a drawback, I didn't want to give anyone the chance to think of me as different or incapable. I hoped to be treated like everyone else and not be excused from any tasks. Growing up with three older brothers, I knew the feeling of being excluded from sports and games because I was a girl; I would be damned if that would happen in my eagle work. Certainly, a latent fear of heights would not be acceptable for a raptor biologist who needed to scale trees and cliffs and, in this case, towers, to access the birds.

Inside my head, I repeated the mantra *step by step* as I planted one foot on the first one, slowly pulling my body up to meet it. My hands gripped the steel so hard my knuckles turned white, my palms exuding slippery sweat. I wished I had worn gloves so they could stick to the metal. The next foot came up, my knees shaking now, and I stopped to be sure it was firm in its foothold. I looked down, a mistake, but I was only a few feet off the ground, just enough to know I shouldn't look down again. I had to divorce my mind from my body, had to think only of the movement of my limbs and not of the consequences of a fall. One more step, then another, and I started to count. The numbers took my brain to another place as I wondered how many more there were, how many steps equaled the height of the tower, how many more body lifts it would take to reach the platform. I could hear Jim's voice above me, but I was too distraught to make out what he was saying. Finally, his quiet and encouraging words "almost there" reached me as I moved upwards, and I knew then I would make it. My innate stubbornness came to my rescue, and somehow, I reached the top without revealing my fear to Jim. My thumping heart and wet palms were invisible, a blessing. Or maybe he knew all along and was just too generous to call me out. I will never know.

---

Several days before the birds arrived, Jim, Willard, and I accessed our inner eagle and built a nest on the platform. We tried, to our best human abilities, to make it look like the nest they would be coming from in Wisconsin. Luckily, Jim had traveled to Canada and Alaska to study Peregrines so had seen many wild eagle nests. He knew those of Bald

Using a stepladder to reach the tower's metal footholds, the author climbs the pole up to the platform to build a nest before the eagles arrive. *Photos courtesy of Larry Morris*

Eagles differ from Golden Eagles by being a bit messier. The sticks are more loosely fit and look more haphazard. It would be stressful enough for the young eagles to be taken from their home, confined to a crate, flown in a jet to a state many miles away, and then plopped into a strange nest. We needed to make it look as familiar as possible. I was aware, looking down at the bare wooden platform before me, how much we were asking of these birds. Not only were they leaving their home and parents, but they were being asked to become independent flyers, predators, and breeding adults without the benefit of having been taught these skills by the adults of their species.

Together, the three of us formed an assembly line. Eager to remain on *terra firma*, I volunteered to collect dead branches from the nearby woods. Willard bundled these with a rope, and Jim hauled them up to the platform to assemble them. Using sticks of varying sizes, positioning the larger ones in a four-foot circle, then interspersing the smaller ones in between the cracks for stability, produced the desired effect. After the task was completed in a few hours, all three of us held more regard for parent eagles who had to collect their sticks one at a time, fly with them in their beaks, and then weave them into a structure that could withstand both wind and the antics of juvenile eagles. I was pleased we could even come close to mimicking their herculean efforts.

Thankfully, before the eagles arrived, scaffolding was erected alongside the platform to hold the canvas blind or hide from which I would be able to access the nest. The metal frame would also serve as a ladder leading to the platform. In the end, I would climb up and down that ladder many times a day until it eventually became second nature. Once I learned to never look down, my fear morphed from numbing terror to mindful caution, laced with a healthy respect for heights. Although my terror never returned after that first ascent, I worried there would be many more tests I would have to pass.

## — Chapter 9 —

# Welcome Party

Uncertainty and panic engulfed me, blurring the six weeks preceding the eaglets' arrival. Whenever I was having anything approaching a normal day, struggling to live in the present, images of the unfamiliar terrain ahead launched me into the future. Tomorrow was always arriving before I was ready and had my feet solidly underneath me. I looked forward to and dreaded, in equal measure, the start of the project, when the fantasizing would stop, and reality would begin.

Bald Eagles usually hatch in late April through early May in the Great Lakes region, once the lakes and rivers experience ice-out, the daytime temperatures rise above freezing, and the sun's strength works its magic. After the long winter, parent eagles can now find enough food to feed their ravenous offspring.

But I had no idea of the exact date the climbers in Wisconsin, hired by the Fish and Wildlife Service to band eagles, would select two nestlings to send to New York. Usually, the best time for banding is when the young are between six to seven weeks old, the age when they are feathered out but still too young to jump from the nest. This would indicate a mid-June arrival date, barring any bad weather delaying the climb.

I had been told to be ready to head to the Montezuma refuge as soon as the plane carrying the eagles was on its way east. All I could do at this point was await word from the Wisconsin banders, with little to occupy my time but packing and repacking my camping gear for a summer in the wild. Since I had no idea what equipment I needed, what clothes I should bring, or what I was heading into, every outdoor item I owned was strewn throughout my apartment, ready for me to decide its fate.

## 1976

The phone rang with the news on June 18, 1976. Two eaglets had been secured and were on their way to New York. It was finally happening. The apprehension I had been feeling, the disquiet roiling inside me, the restless nights and unsettling daydreams about the adventure I was embarking on, were about to be over, or so I thought. It turned out these emotions were just smoldering, like ashes primed to reignite without warning.

---

As I came to find out from Tom Cade later, the Wisconsin climbers had a sudden change of mind, deciding to take two birds from the same three-bird nest, leaving only one to be raised by the parents. We never knew if this was a decision based on the condition of the birds in the nest (did it look like the parents were having trouble finding food for the chicks?), the weather (did the climbers decide they should stop at one nest because an impending storm could preclude another climb?) or something else entirely. We only could say it was not the usual practice since scientists didn't know if it was in the best interest of the species to take birds from the same nest.

When breeding animals in captivity, biologists try to separate siblings to avoid inbreeding, or combining similar DNA, hoping to promote diversity and increase the likelihood of better adaptation. But in the wild, eagles might not be reading the same rulebook, and little was known about mating among siblings or family members. As it turned out, our assumptions about sibling relationships proved to be unfounded, requiring us to reevaluate the manual.

---

Once I received the news of the eagles' arrival, I went into overdrive. Their plane was expected at the Syracuse airport by late morning. I threw my camping gear and several bags of clothes and food into the trunk of my car. Locking my apartment, I dropped a key off with my neighbor so she could water my plants over the next few months, loaded Max and Moses into the backseat, and drove north to meet the eagles. Having no conception of what I was heading towards, I tried to keep my misgivings at bay as I focused on the shoreline of Cayuga Lake drifting past my window.

I was the first to arrive, but a small crowd trickled in behind me onto Clark's Ridge—my home for the next few months—to greet the nine-week-old eagles. Cornell raptor biologists Tom Cade, Jim Weaver, and Willard Heck; videographers Lance Wisniewski and Pat Faust from Innervision Media Systems in Syracuse, hired by New York's DEC to document the project for the local PBS station; and more than a dozen refuge and state personnel began milling about as we waited for the action to begin. Max and Moses inserted themselves into the middle of the group, confused about what was happening but hoping for attention. As time went on, they grew bored and lay down in the shade of the nest platform, allowing me a breath of relief that they were out of the way.

Everyone waited in anticipation, voicing curiosity as well as concern. None of us knew what to expect in this ground-breaking project. I worried that the audience would look to me, the on-site researcher, to take charge once the eagles arrived. Since I had no idea what I was doing, had never even handled an eagle before, I relied on others for direction, support, and courage. Although at some point I was to take over the care of these valuable birds, that time hadn't yet arrived. On that day I knew my survival, as well as the eagles', depended on the Peregrine team having my back.

In an hour, which seemed interminable, the Fish and Wildlife vehicle that had met the plane at the airport rattled up the dirt road toward the nest tower with their charges in tow. As the truck approached, I pictured majestic eagles sitting proudly in their crates, heads held high, ready to impress all of us with their grandeur. The actual scene inside the boxes was probably quite different, evidenced by the clamorous sounds of frantic eaglets banging their wings against the sides of their crates, announcing their arrival.

Our first order of business was to get the birds into their new nest as quickly as possible before they injured themselves. Once the truck had backed up to the tower, Jim and Willard unloaded the two crates onto the ground. Cade, Jim, and I climbed up to the nest platform to help receive the birds when they arrived at the top. By now, my worries about the climb were diluted by my worries about what was going to happen once the eagles were released onto the platform. With the crowd and the Peregrine team around, I had help, but what would happen when they all left? I tried to stay in the moment, relegating my concerns to the back of my brain until I needed to deal with them.

Willard remained on the ground to secure ropes around the crates below the tower. The eagles continued to thrash, uttering low guttural clucking alarm cries. Jim hauled them up, one at a time. Once the crates were on the platform, Cade and I untied the ropes. Lance and Pat had joined us by then with their video cameras whirring away to capture the birds' ascent into their new nest.

The crowd at the bottom craned their necks to see what was happening on top. Even if anyone had been curious enough to attempt the climb—and I certainly could empathize with those who weren't—there was barely enough room for the five of us, the crates, and the eagles once we released them.

The eaglets themselves were light—no more than a few pounds apiece—but the crates were solid. Once on the platform, we used a crowbar to pry open the lids. With beaks clacking and wings flailing, both birds leapt onto the nest and stood like statues. Their stares seemed ominous, but an eagle's expression always appears to reflect suspicion, so it is easy to misinterpret their emotions. Their fierce glare is due to their unique eyebrows, created by a supraorbital ridge believed to protect their eyes from the sun. Since birds have no facial muscles, it is difficult to read what they are trying to convey based only on their expressions. I would have understood, however, if they had been distrustful of us. Their first exposure to humans had resulted in being removed from their parents and nest and flown halfway across the country in a wooden box, a traumatic experience for any animal.

As I watched them teeter on the rim of the nest, part of me hoped for a sign of recognition, an indication that they knew I would take care of them, would do everything I could to do right by them, but at that moment I knew that there would always be a certain gulf between us. They had been born wild and needed to remain that way. Unlike my dogs who were sleeping below the tower at that moment, content and safe in their dependence on me, these eagles must remain aloof and wary of humans for their own survival.

We examined the birds closely as they hopped about the nest, to be sure they hadn't sustained any injuries during their trip in the crate. Satisfied they seemed healthy and in good shape, we climbed down from the tower, leaving them to settle into their new home.

With the action over, the crowd dispersed, returning to their cars. It didn't hit me until the onlookers had left, Cade and his crew had pulled out in their vehicles, and the photographers had taken off for the B&B in town, that I was now alone. Besides Jim and Willard, who promised they'd return in the morning to check on me, nobody had said goodbye, wished me luck, or even acknowledged that they were entrusting the eagles to my care. Looking back, perhaps I should have wondered why the observers on the hilltop that day acted as if the publicity surrounding the eagles' arrival was of equal importance to keeping them safe and secure. But my mind was consumed by other matters. The pressure of keeping these birds alive, of playing Mother Eagle, of being responsible for the national symbol—all of this was weighing on me as I watched the last group disappear down the road.

Luckily, I had Max and Moses for company, so I didn't feel completely abandoned. They became avid listeners as I talked to them aloud, if only to hear a human voice. Although I don't remember my exact words to the dogs, I imagine "What now?" might have been part of our conversation. I forced myself not to dwell on being isolated on this hillside for the next four months and instead focused on my biggest conundrum of the day: how to provide the eagles with their first meal.

## — Chapter 10 —

# First Supper

That evening, after everyone had left the ridge, I approached the eagles' dinner with a bit of trepidation, unsure how they would react to not having their parents deliver their food. I had been so worried about this I had taken precautions to set everything up ahead of time. I knew I would be panicked if I had to organize their feeding at the last minute.

Two days before arrival day, I had driven the forty miles from my Ithaca apartment to the refuge, partly to look over my future campsite before my attention was diverted to the birds, but also to figure out the logistics of feeding.

As parent eagles spend much of their day searching for food for hungry nestlings, I knew my days would be similarly occupied. When designing the nest tower, Jim and his crew had taken many steps to make the feeding process easier for me. The platform was open on three sides, and off the fourth sat a small plywood hide, or blind. By climbing up the supporting scaffolding underneath, I could pull myself into the hide, and haul up a bucket of food with a rope, all without the birds seeing me.

It is crucial that eagles, like many raptors, don't identify their food source with a person or they might imprint on you. They begin to think of themselves as human, lose all fear, and eventually can't recognize other eagles as their own species. This upsets their normal breeding behavior since they also won't recognize displays and calls that lure them into being receptive to the opposite sex. We knew our birds had seen their parents during the weeks before they were taken from the nest, so if that image could stay fixed in their minds, the hope was that they would think and act like an eagle. I often thought of Shirley, one of the cheetahs at the zoo, who was so imprinted on humans that she couldn't breed. Although her friendliness and dog-like behavior had endeared her to the zoo staff, it made me sad that she wasn't able to accept her identity as a cheetah. This memory made me doubly conscious of keeping my identification as a food source a secret.

I rigged up a long stick with a barbeque fork taped to the end, which I could use to pitch food onto the nest through a hole in the canvas wall facing the platform. They might see the fork as their mother substitute, but at least they wouldn't attach their feeding to a human.

The refuge staff had shown me the spillway at the edge of the dike road where carp from the marsh would congregate and become trapped. An invasive species that reproduces quickly, carp can damage habitat and outcompete native fish. I was given permission to catch as many as needed for the eagles, the rest being sold by local fishermen for fish sticks and other products to keep the population in the pools from getting too large. This plentiful and accessible food source would become a mainstay of the eagles' diet.

The carp were often over two feet long and weighed around eight pounds, so one fish would last for a day. Predicting the birds would be hungry as soon as they arrived, I stockpiled some food for them ahead of time. Using a long-handled net, I grabbed a few of the biggest carp trapped in the spillway and plopped them into a lidded plastic bin I could drive back up to the hillside. Once there, I transferred them into a thirty-gallon trash can filled with water. Carp are resilient and can live a long time in a confined space, so I thought it would be easier to keep several live fish in the bin at my campsite to avoid daily trips. I tried to catch enough fish to last a few days since the trip from the hillside down to the marsh took fifteen minutes in the International Scout—a Jeep-like SUV produced by International Harvester until 1980 and loaned to me for the summer by DEC.

My next challenge was to kill the carp. Parent eagles do this for the young, but in their absence, it was up to me. It seemed wise to practice my technique before the eagles arrived since I had a hunch I wouldn't be a natural at it. These fish are big and tough with thick scales and large heads. The only quick and humane way to kill them was to hit them on the head in just the right place with a heavy instrument. I tried all manner of killing sticks until I found the best one: a heavy axe handle borrowed from the refuge station.

I am not sure I had ever killed an animal before—much less bludgeoned one to death—even when my dog had brought home a mortally wounded bunny. In fact, this was the worst part of my job and only accomplished when I convinced myself it was for the good of the birds, and that these fish were doomed anyway, with or without my assistance. In the months that followed, my clothes became perpetually blood-stained from this ritual, wreaking of a fishy smell that wouldn't come out despite repeated washing. I noted with some amusement that visitors to the hilltop kept their distance from me if they arrived on a day of fishing.

---

I stared up at the tower that first evening, planning my next task: serving dinner to two hungry orphaned eaglets. I grabbed a carp out of the barrel using a gaff hook. The large fish wiggled about on the ground, but practicing my killing technique had paid off, allowing me to dispatch the animal quickly. That first time was horrific, filling me with guilt and disgust at ending the carp's life, and hope that it hadn't suffered. But after doing this twice a day for months, I eventually became inured to it.

Putting the carp into a plastic bucket and fastening a long rope to its handle, I began the climb up to the platform, looping the rope around my wrist. I had left the fork up there before the eagles arrived, so it was ready for action when I maneuvered my body into the small blind next to the nest. I hauled the fish bucket up after me, trying to be as silent as possible. I needn't have bothered since the birds remained calm, unaware someone was on the other side of the canvas wall. Securing the carp on the fork, I pushed it through the hole and flipped it over the lip of the nest into the cavity.

I expected the eaglets to inspect the fish, to consider what this might mean, to wonder how it had arrived, but they were not contemplative when it came to food. Both jumped on the fish at once and, starting at the puncture wound created by the gaff, tried to rip through the thick scales to the flesh below. Their feeding techniques would improve with age once they learned to coordinate their talons and beaks to secure and then tear apart their prey. But on this day, their first experience with independence from parents, they forgot about their talons and instead shook the fish, dropped it, and then shook it again, as if hoping the meat would drop out. Although their efficiency needed work, the power of their beaks and talons was impressive and made me aware how dangerous these weapons could be if directed towards me. Even though they were only nine weeks of age, the birds deserved my respect as does every wild creature. This belief has always guided me.

I sat motionless in the blind, willing the eagles to remain unaware of my presence after the carp appeared before them. While watching them devour their first meal, I began to register I was finally here, caring for young birds whose importance to the world clearly outranked my own. A pounding in my chest reverberated through my body, my emotions threatening to spill over. Moments like this came along whenever I found myself alone with the birds, whenever I had time to reflect on the surreal position in which I found myself. One minute I was watching predators devouring fish, the next seeing the nation's bird struggle to survive.

Leaving the eaglets to tear apart their meal, I returned to the ground where the more mundane tasks of digging a latrine, setting up my tent, and putting my belongings where I could find them again awaited me. Although I had never been a Girl Scout and my parents hadn't taken us camping when we were young, I had done a lot of backpacking with friends when I was in college. Since all of us had wallets running on empty, camping was the vacation choice that made sense. We could be outdoors—hiking, canoeing, and fishing—and not worry about lodging and restaurant expenses. Gradually, I had learned the basics of pitching a tent and making a camp feel as hospitable as possible, skills I was grateful for as I created my new summer home.

As the sun was disappearing below the horizon, I was anxious to get started, not wanting to be groping my way around in the dark. I found an ideal place for a latrine, hidden within a thick grove of trees but with the bonus of a panoramic view of the marsh. A perfect location for some deep thinking! My tent site was trickier: since there was so little level ground on the hillside, I had to compromise by sleeping in an uphill direction. I was too tired to care that first night so decided to tackle the problem when I had more energy.

Since I knew my tent would be my only shelter for several months, I had treated myself to a six-person canvas model, guaranteed to protect me and my possessions from anything short of a monsoon. Inside I would have to fit all my food—no bears had been reported in the area, but raccoons might have enjoyed my goodies—as well as a few changes of clothes, sleeping bag, books (from natural history to novels to memoirs by biologists—some for entertainment, others for inspiration), spotting scope, binoculars, camera, camping chair, and of course my two large dogs. All should be safe and secure, or so I convinced myself, as I turned in for my first night on Clark's Ridge.

As I climbed into my sleeping bag, after polishing off the two sandwiches I had packed that morning, I realized how vulnerable I was. Besides the eaglets, high above me in their nest, and the dogs, there was nobody within miles of the hilltop. If I had been camping in the wilderness, this would have been true as well, but at least there I might have been reassured by the unlikelihood anyone could intrude on my solitude. Here, dirt roads led to my campsite, and I had no way of knowing if my existence would be discovered. Max and Moses may have felt the same, or may have just known I needed comforting, because they lay down on either side of my sleeping bag, their warm bodies eventually allowing me to drift off.

## — Chapter 11 —

# A Rocky Start

By 3 a.m., a monsoon had arrived. I was woken from a sleep so deep it must have been making up for the past weeks of anxiety-ridden nights. Cold water was dripping onto my face, infiltrating my dreams until it finally broke through into my consciousness. The tent was listing hard to one side, the impermeable canvas leaking rain as if there were no barrier at all. Glancing at the luminous dial on my watch, I knew I had miles to go before the night would end. Too groggy to deal with this, I dragged my sleeping bag to a drier spot and fell into a fitful slumber, waiting for first light to reveal the damage.

When my eyes opened again, the floor of the tent held an inch of water, the fly had blown into the trees, and all my precious belongings, except my optical equipment (safe inside a plastic tub), were soaked. The perishable food stored in the cooler remained untouched, but the dry goods were ruined. My clothes, sleeping bag, and books were soaked. Even the dogs were wet from lying on the bottom of the flooded tent, but at least they were calm, unlike their owner.

I wasn't sure whether to cry or scream at the heavens. Although there was nobody around to hear me, the desperation I yearned to spew forth stayed bottled up inside me. Once it found an opening, I feared it would forever remain unleashed. Instead, I focused on the immediate: I needed help, and a new tent.

Jim and Willard were due to return to the refuge before noon to help me band the eaglets.

One option was to wait until the two men could observe my dire straits for themselves. But for them to bring a new tent, I needed to contact them before they left Ithaca for Montezuma, almost an hour's drive away. My only means of communication was to drive down to the refuge station—three miles by car—and call from there. With my soaked body oozing water over the front seat and the bedraggled dogs following suit in the back, I pointed the Scout down the hill. Holding back my tears on the phone when Jim answered was hard, but I knew it was essential. This was just the beginning of the challenges that awaited me, and I didn't want everyone seeing me fall apart over a little rainstorm. Jim's voice, which always seemed to grow calmer as those around him became more agitated, was true to form. He settled me down with assurances he and Willard would be coming to my rescue as soon as he could get his hands on dry equipment.

Two hours later, Jim's pickup arrived with a new tent, food, and waterproof containers and tarps; my wet clothes had dried; and life was restored to normal again. I was able to regroup with a fresh outlook, convinced that I could do this after all.

We spent the rest of the day repairing the canvas blind, which had also suffered leaks during the storm. The wind whipping across the marsh moved into high gear once it hit the hillside, so we needed to tie down everything more securely than we had thought. It was probably lucky that the rain happened on the first night so we could shore up the site early on—although I didn't feel so lucky at the time.

We had made another miscalculation that might have resulted in the eagles' undoing: we hadn't expected the young arrivals to be as mature and feathered out as they were. Weather issues had delayed the climbers in Wisconsin from taking the birds from their nest site on schedule, so they were at least a week older than we had thought. Assuming their flight feathers would be growing in but not fully developed, we were all surprised when the eaglets emerged from their crates. There is a big difference between a seven-week-old and nine-week-old eagle. Ideally, the less developed birds would have had a couple of weeks to spend on the platform, gain strength, and bond with their nest site. Instead, these two eaglets, estimated to be nine weeks of age based on their plumage development, were close to being able to fly, or at least leap, off the platform into the great beyond.

Wild eagles usually nest in mature pines or other conifers, or tall hardwoods when available, where solid branches support the nest and

The tower with one barred side in place to prevent the eaglets from falling off; two more sides to go. *Photo by Cornell University*

offer roosting spots once the young are big enough to leave the security of the cavity. In our case, there would be no tree, just a human constructed nest. Although we had tried to replicate the one they had left in Wisconsin, we could provide only a one-foot perimeter between the nest rim and the edge of the platform, hardly enough to keep a feisty eagle, eager to practice its flight skills, from the thirty-five-foot drop to the ground below.

As dusk descends, an eagle's energy level rises, for this is the time of day to search for food in the wild. Their leaps around the nest rim and the jumping jacks performed as they flexed their wing muscles would surely have carried these birds off the platform, making them vulnerable to predators. We were lucky to have foreseen the problem before it happened, thanks to the storm alerting us to the potential danger and allowing us to dodge catastrophe. To keep the pair safe, we constructed three wooden frames with vertical aluminum bars that we hoisted onto the platform to enclose the three open sides. The blind would block off the fourth side. At least this way calisthenics class could be held without risk.

---

After everything was battened down, Jim and Willard headed home. I was warm and dry, looking forward to a can of Dinty Moore beef stew—the quintessential camping cuisine in the days before freeze-dried gourmet meals were invented—cooked on my Coleman propane camp stove. The last twenty-four hours had seemed like a false start to my adventure, disappointing after all the hype and preparation, but providing a fresh beginning once I had readjusted. After dark, I sat outside the tent, scanning the clear sky for shooting stars.

The perfect view of the Milky Way reminded me more of the western sky I had seen on summer sojourns in the Rockies, undiluted by city lights and nighttime traffic, giving it a purity that made me want to fold myself into it. Relief that we had discovered our oversights early enough to correct them swept over me. My anxiety about the task ahead hadn't disappeared but had dissipated a bit. Whenever panic overtook me, it was usually during those moments when I saw myself as alone in this endeavor. I imagined all decisions rested with me and if mistakes were made, they would be all mine. Awareness of my own limits and lack of experience would become my hardest truth in the coming weeks. This day of missteps and regrouping had been a blessing. Despite the ruined tent, wet equipment, and the vulnerability of the eagles, I had swallowed my pride and asked for help. I wasn't the sole proprietor of this project unless I wanted to be, and for now, I still had much to learn before I was ready to handle everything by myself.

The dogs and I slept well that night, despite Max and Moses waking often at the sound of raccoons nosing around outside the tent. The predator guards and bars around the nest were in place, easing my mind, but I wondered how many other surprises awaited me.

## — Chapter 12 —

# Flying Solo

Once the young eagles had settled into their new home, many emotions threatened to unseat me; the one I was least prepared for was loneliness. I would go days without seeing a soul unless I called the Peregrine crew for help with the birds, or a refuge employee stopped by to bring me messages. I had never spent long periods so totally alone and wasn't sure how to deal with these new feelings. Eventually, I found the best cure for such melancholy was a daily schedule with each minute crammed with activity. The eaglets kept me to this since their feeding times had to be consistent and their behavior chronicled as often as possible. My research depended on evaluating every aspect of the hacking technique. A study of the young eagles wouldn't be deemed legitimate science if the results relied on my subjective interpretation.

It wasn't enough to record how impressed I was when they hopped about their platform, how proud I was that they were strengthening their wing muscles each day. Nobody cared how I *felt*. Instead, I had to time their exercise regimens, breaking them into discrete categories like wing-flexing and leaping in air. All behavior had to be quantified and statistically analyzed to assess whether it was within the parameters of normalcy. The main issue was simple: Were we affecting the birds' behavior by taking them from their parents and releasing them to the wild through hacking? Other studies of wild eagles rearing their young allowed for comparisons to be made, assuming my records of the hacked young were thorough and therefore valid. A researcher's greatest worry in any reintroduction project is the alteration of a species' behavior, something to be avoided at all costs. The hope was to have the animal blend into its natural habitat as if it had been raised there by its parents.

To study this, I had to do something that went against everything that came naturally to me. I had to separate myself from the eagles emotionally and think of them as research subjects rather than pets. To make this even remotely possible, I decided not to give them names, which would have drawn me even closer to them as individuals. There is always a connotation associated with a name, either to a known person, a characteristic, or special quality. My hope was to remain as detached as possible by identifying them with the state they were from and their relative size. In 1976, for example, the Wisconsin siblings were W1 and W2. W1 was larger and therefore probably female, and W2, the smaller male. I had to fight my urge to adopt them as my own birds. I was lucky to have Max and Moses there to satisfy my need for animal affection.

As soon as the sky over the marsh lightened, my days would begin. My favorite part about studying birds is that most—owls and a few others being the exception—are diurnal, or active during the day. They usually sleep until just before dawn, unlike mammals who are nocturnal or most active at night. I wanted to be ready to watch the eagles at sunup while it was still cool. As the sun rose higher in the sky, the heat made their energy flag.

My first task was to climb the scaffolding to the blind with a fish or the special "roadkill of the day" inside my bucket. After feeding the birds with the fork, I sat back out of sight and watched, taking notes on everything they did. I had made a list of the possible behaviors they engaged in—eating, preening, beak wiping, vocalizing, fighting with each other, flapping, and so on—and then timed and counted to create a timeline of their activities. They couldn't see me but came to connect the sound of my steps on the ladder to the arrival of food. They would stand on the edge of the nest and emit an ear-splitting caw-like cry as they anticipated the emergence of the fork. This was the extent of any positive feedback they gave me. I had to accept that my affection for them was not reciprocated.

The pair was active for only a few hours in the morning and evening. During the heat of the day, they spent their time napping, panting when it was hot (they didn't require water since their food provided necessary liquid), and looking beyond the tower. Fascinated by their surroundings, they were most intrigued by other birds overhead and the occasional plane. When excited, they would clack their beaks and sometimes flap their wings, jumping up and down in place. Often, I would laugh out loud at some of their antics, so similar were they to human toddlers when first exposed to something new and exciting.

This was hardly the romantic research I had envisioned when I started out. Being three feet away from an eagle nest was a rare experience, to be sure, but was more a lesson in monotony, occasionally peppered with wonder. My naivete disallowed any concept of how much down time there would be and how methodical most research was. In truth, the drama of field research is a fairly delusional concept. I had read of the adventures of George Schaller, Dian Fossey, Jane Goodall, Barry Lopez, and others as they tracked and observed lions, gorillas, and chimps, followed wolf packs or killer whale pods, or tagged grizzlies. I couldn't wait to join the club of wildlife biologists living on the edge of discovery and inspiration. Goodall provided reason for my conflicting thoughts about research when she wrote about the chimps she studied: "When science says you have to be coldly objective [and] you can't have empathy, they're completely wrong. . . . If you have empathy with your subject you are more likely to understand complex behaviour." Whenever I felt guilty about abandoning my objectivity and becoming too attached to the eaglets, I would think of her words. For me to succeed as a research biologist, I needed to relate animal behavior to something familiar, to give it context to understand it.

Luckily, I am an avid reader, so while the eagles slept and my other chores were done, I escaped into worlds as different from my current one as possible. The drawing rooms and moors of nineteenth-century England transported me far from my hilltop home, as I retreated into Austen and the Brontes on many a quiet afternoon or evening. Despite my quest to enter another realm, however, the flatness of the marshes with their endless swaths of cattails paralleled the heather and cotton grasses of the Yorkshire moors I was reading about. I often wonder if I would have sought solace in these literary works if my iPad or iPhone were available; I am quite glad they hadn't yet been invented or I might have found myself streaming forgettable movies for escapist entertainment.

As the temperature rose by mid-morning and the eagles' activity slowed down, my own energy flagged as well, for it became sweltering inside the canvas blind. I descended the ladder, joints aching from sitting cross-legged for five hours. To stretch my muscles, I took walks with the dogs down the dirt road or in the nearby woods. With binoculars, I often explored the areas abutting the hillside with the hope of seeing indigo buntings and red-headed woodpeckers perched on the fence posts surrounding the plowed fields. Sometimes, my birdwatching extended into the wetlands below the hill where I could sometimes catch herons feeding in the shallows or, if I was lucky, an occasional rail running stealthily amid the vegetative edges of the small pools. I also used this time to take the Scout out to find food for the birds, searching the farm roads for roadkill.

Bald Eagles eat a variety of food and are not just fish eaters like Ospreys. They will often feed on carrion when it's available, leading to their reputation as scavengers as well as predators. The Scout and I—with Max and Moses riding shotgun—made daily excursions on the roads surrounding the hillside to look for roadkill that might be appetizing. In the back of the Scout, I had placed a shovel and a Rubbermaid bucket so I could be prepared when a carcass appeared. Eventually, I became proficient at evaluating a roadkill's worthiness. If I saw flies buzzing around it, it wasn't fresh enough. My day was made if I could find a rabbit, but these were rare since they were adept at avoiding cars and were not active at night. Having once had rabbits as pets, I lamented their last journey across the road, their ultimate enemy not a fox or coyote but a hunk of fast-moving metal. Yet I also rejoiced in knowing my young birds would enjoy their favorite meal.

Other animals that appealed to the eagles were squirrels (the most common roadkill), muskrats (which smelled nauseating, but luckily most birds have poor senses of smell), and woodchucks (plentiful but unappetizing and usually ignored). I put my foot down at skunks. I became so obsessed with looking for carcasses on the roadsides that I did it even when I wasn't looking for eagle food. To this day, while others might look away from an animal killed on the highway, I crane my neck to identify it, often wondering about its life before it was hit.

Roadkill patrol was time-consuming, so I asked the refuge staff to keep their eyes out when they commuted to work. They knew the project was top secret, but if they found suitable roadkill, they would bring it to me. One staff member, Ann Harrison, went so far as to give me a snake she had found. I felt badly the eaglets didn't show any interest in her donation. Fortunately for everyone, carp remained the mainstay of the diet.

Whenever the garbage can called for a fresh supply of fish, I drove down to the spillway to catch more carp. The dogs enjoyed these excursions, and I let them run alongside the dike road and in the marsh areas while I netted fish for the eagles' meals. On rainy days, rare during that summer, I sat in my tent and edited my research notes. My illegible chicken scratch, which substituted for handwriting while crouched in the blind, needed regular deciphering.

---

Once a week, I treated myself to a trip into nearby Seneca Falls to buy groceries and bags of ice for my cooler and to refill my water containers. Although nervous about leaving the eagles unattended for more than a couple of hours, these sporadic expeditions became precious to me. Occasionally I was even able to shower and do laundry—a true luxury—thanks to the generosity of Ann Harrison, donor of the snake, who had an apartment in town and lent me a key. I sought out conversations with everyone from grocery clerks to gas station attendants, so starved was I for human contact.

One of my favorite haunts was the supermarket bakery where I would treat myself to a cheese Danish. The lady behind the glass case came to know me, even reached for the Danish before I ordered, and soon became a friend who asked me about the eagles and my life, a simple interaction that reminded me I was still a member of the human race. I had never eaten a cheese Danish before and have not had one since that summer. Since I couldn't buy any food that needed cooking or refrigeration, a customer walking through the market might have wondered about the bizarre woman who stared longingly at the potatoes, broccoli, and lettuce.

I could never completely relax or linger when in town, so worried was I about abandoning my charges. There was no gate, no bars, no guard blocking my exit. No, instead there was my conscience telling me my job was to care for those eagles, and I was skipping out. Although leaving the hilltop occasionally was essential for my mental health, I always felt guilty doing it.

When I returned after my weekly trip into town, I approached the tower with a pit of dread in my stomach. I placed undue pressure on myself to be the birds' protector and allowed my imagination to run wild when they were out of my sight. Returning to the nest tower, I was haunted by visions of intruders, accidents, escapes, and other atrocities until I could see that all was well. Years later, when my children were infants, this same fear haunted me as they lay sleeping in their cribs. Whatever maternal instinct that kicked in to induce such worry was the same as the one I had for the eaglets.

Ironically, in 1848, Seneca Falls had been the birthplace of the women's rights convention, yet I was so immersed in my own struggles to prove myself that I never recognized the impact the town and its history had on my life, and that of every American woman, until many years later.

---

Although my campsite was hard to locate—the unmarked farm roads forming a maze as they followed the edges of cornfields—that didn't stop the occasional visitor from wandering up the road uninvited. I was lucky to have my large German shepherd with me, with his even larger bark. Max hated intruders who were on foot and tore down the road after them until I called him back. Usually, these surprise guests were hunters or hikers who had lost their way. They were relieved when I gave them directions back to the main road, and even more relieved that I was holding Max by the collar while we chatted. The dog's presence was registering with them, making me feel better in case there were future encounters. Moses, my collie mix, was a lost cause, on the other hand, when it came to strangers. He ran towards everyone with tail wagging, determined to become their lifelong friend.

On a few occasions my solitude was disturbed by acquaintances who decided to take advantage of my need for company. One of these visitors was a former teacher, newly divorced, who arrived with an elaborate picnic, complete with plastic wine glasses and candlesticks. How amazing that a professor had recognized me and been willing to come all the way up here to spend time with me! I was flattered by his generosity and interest in my project, and grateful for the break from my usual canned dinner fare. At first our conversation centered on my eagle work but somewhere along the line things shifted. It dawned on me that he had been planning to share my tent that night. Betrayed and embarrassed, I asked myself how I had misread the signals. By considering me fair game—a woman alone in the middle of nowhere desperate for company—he had denigrated the importance of my research and my overall worth. I knew I needed to be more vigilant. Max couldn't rescue me from every intruder. As smart as he was, he had as tough a time reading the signals as I did.

Days would go by when my infrequent conversations with anyone were limited to the subject of eagles. Visits from Jim and Willard lifted my spirits but became more intermittent when they had to travel in late June and July, supporting the various Peregrine hacking sites and attendants. However, whenever I had a problem that I couldn't solve or when I needed assistance with banding or equipment, Jim's truck would appear like a rescue vehicle. I looked forward to his arrival, for it was then that I learned about banding, radio telemetry, raptor behavior, and the art of improvisation. If something was needed and nobody had yet invented a tool or apparatus for it, Jim would rub two sticks together and make it happen.

Larry, an important part of my life by then, was spending his summer in New Brunswick, Canada, where he was running an environmental camp for the children of local loggers, fishermen, and road construction workers. He too was isolated without phone service or even a post office. Of course, this jumbled my feelings; not only did I have someone to miss, but we couldn't even see each other or communicate during this time. The difference was that he was surrounded by campers and fellow counselors, whereas I was alone. Envisioning our contrasting experiences brought forth unbidden envy. Although I tried to put this out of my mind and focus on my work, I wasn't always successful.

Other friends and colleagues from Ithaca were also doing their summer fieldwork, often in other states, and couldn't come up to Montezuma to visit. In the nick of time, one day when I was feeling particularly sorry for myself, my good friend PK, who was doing fisheries work in the Adirondacks most of the summer, appeared at my campsite, unannounced since she couldn't contact me. Hidden in the large cooler she pulled from her car was a delicious home-cooked dinner and, best of all, she brought hours of conversation, making me smile for days afterwards. She had no idea what that meant at the time.

Solitude is a place hard to fathom until you are inside it. That summer, I understood why it is used as a punishment in prison. I felt ungrateful during these low moments, for hadn't I asked to have this opportunity, hadn't I wanted to be a wildlife researcher? Then I would snap myself out of my funk by thinking how lucky I was to be on this hillside. I would look at the two young birds in their nest, the embodiment of the conservation movement of the 1970s, and know I had to pull my own weight if the three of us were going to succeed. Whenever the doubts seeped into my soul, I tried to remember the pilgrimage I had undertaken to reach this point.

## — Chapter 13 —

# Search and Rescue

My peaceful, predictable days of observing and feeding the eaglets and hunting for roadkill were to be short-lived. As the birds gained strength and approached their fledging or flight stage in the next couple of weeks, I needed a way to track them once they'd left the nest. This is when my lack of Eagle Mother skills would become most conspicuous.

Once a young eagle fledges for the first time, it often descends to the ground like an out-of-control helicopter. The skills to nail its landing on a nearby tree branch have yet to be mastered. An immature eagle's large size serves as a disadvantage during this period. By the time it is ready to fly, it can exceed ten pounds and require so much lift to become airborne that first flights are not a pretty sight. In the wild, parent birds monitor the fledglings' movements and feed them on the ground until they have the wing strength to gain altitude. If an eaglet lands in thick undergrowth, this can take several days, especially with no breeze to provide lift.

I would have been helpless in locating the birds without the aid of technology. Equipping them with radio transmitters was my only hope in tracking them and keeping them safe until they were strong enough to fly independently. The conundrum was where to attach the radios on their bodies so the young birds could be followed in the short-term without being burdened by a permanent fixture. Few people had successfully radio-tracked young eagles before, so neither Jim nor I had a model to follow. Eagles molt most of their flight and tail feathers once a year to replace the worn and damaged ones that could interfere with flying. Ideally, putting a radio on any of these feathers would ensure it would be gone in a year, at which time the birds should have left the area and become independent.

We also needed to band the birds with markers to identify them in the future. A USGS aluminum leg band with numbers on it would go on one tarsus or anklebone. While these worked well and could be easily read if the bird was caught or found dead, we needed a way to distinguish our birds from other eagles when they were alive. A second tag had to be visible from a distance so if someone spotted one of our birds in the wild, they would know it had been released in Montezuma.

Jim and I experimented with many different materials and shapes of tags that couldn't be removed during molt yet wouldn't interfere with the birds' flight. The initial solution was a bright yellow Herculite (a vinyl-like fabric) wing streamer, wrapped around the left wing of one bird and the right wing of the other.

Now, for the exciting part: catching the birds and restraining them for banding without being lacerated by their sharp talons or lethal beaks, not to mention whacked by their powerful three-to-four-foot wings. With small birds, you can hold on to their legs with two fingers and their body with the other three while snapping on the band with the opposite hand. But large-bodied predatory birds make this impossible. We needed a solution that would work not just for banding but also would allow me to handle and retrieve the birds in case they escaped. If an eagle fell from the tower, I had to be able to capture it, restrain it, and carry it back up the ladder to the nest without being torn apart.

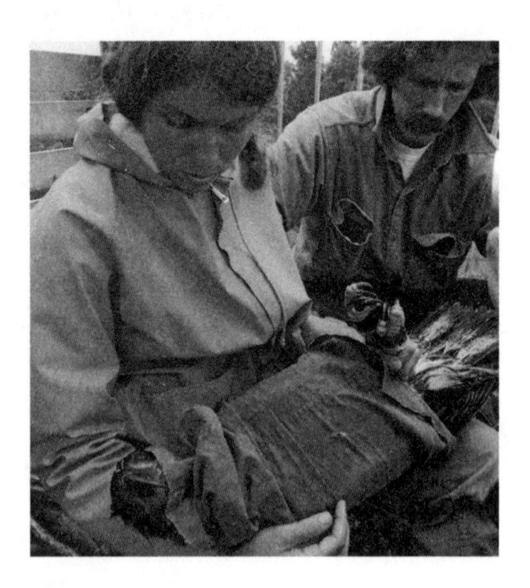

Jim Weaver and the author use a blue-jean sleeve to restrain a 1976 eaglet while they put color tags on its wing and attach its radio transmitter. *Photo by Cornell University*

During my first attempt to secure them, I used an old towel, wrapped it around the eagle's wings, and tried to hold it in place with one hand while holding the talons together with the other. The eagle squirmed, the towel unraveled, allowing him to break loose, scramble to the far end of the platform, and clack his beak at me. This was not going to work. The technique was faulty for many reasons, not the least of which was that it left me without a free hand to fasten the band. To solve this, I considered using a flannel shirt with its buttons fastened so it couldn't unravel, but this wasn't tight enough to pin the eagle's wings to his side. As I looked around for something else, I realized I was wearing the solution.

Blue jeans, it turned out, were the answer. I cut off one of the legs of my blue jeans which tapered a little at one end in the shape of an eagle's torso. Once inside the blue-jean sleeve, the bird's wings were pinned at his side, and only the talons and head projected out either end. Restricted by the tight sleeve, his beak couldn't reach me. By using electrical tape to secure his feet together, I had eliminated any risk from his talons. In this way, I was able to band him, carry him if necessary, and attach a radio to the base of his tail feathers.

One other trick I learned—well known to people who keep caged birds such as canaries, parakeets, or parrots—was that covering a bird's head with a cloth calms them immediately. If they can't see, they remain quiet and immobile.

A patient instructor, Jim never took charge of anything without making sure I understood what he was doing so I could do it myself if he wasn't around. When it was time to secure the aluminum bands to the birds, he explained how eagle bands were different from those of other species and watched me as I attached them to their legs. As much as I tried to be the passive observer, Jim refused to let me. Whenever I handed him the banding equipment, he handed it right back, saying in his calmest voice, "You can do this, Tina. While I hold the leg steady, you slip on the band and fasten it." After the first few tries, I did get the hang of it and felt proud of myself. Although I still looked to him for approval, Jim's expression told me he had confidence I could handle the task at hand.

The radios were trickier since weight had to be a consideration for a large bird of flight. Instead of making the decision himself as to where and how to attach the transmitter, Jim asked for my input. We both knew we were in uncharted territory and brainstorming the right decision was the best way to go. Jim often played around with different equipment back at the Hawk Barn and brought the samples up to the refuge for me to try out: "Which of these do you think will work best? This one's lighter, but this one's battery will last longer." He never let me believe he had all the answers but instead treated me as the caretaker of the birds, often deferring to my opinions, unless they were off base or misinformed. Even though I couldn't have done any of it without his help, I gained confidence and an eventual ability to make decisions on my own.

---

The stronger the birds grew, the more restless they became, like teenagers who are eager to be on their own but don't yet have the survival skills needed to make it. They were starting to develop personalities, their individual characters becoming more pronounced every day. Sometimes I second-guessed my original decision not to give them names to reflect their character traits. I knew, however, that if this program had a future, I should make things simple and easy for other biologists coming behind me to identify the individual birds. So, I swallowed my creativity and sentimentality; W1 and W2 would have to suffice. I was almost

W1 (left) and W2 (right) with streamer tags around their wings and a fish dinner awaiting them in the nest. The larger eaglet (W1) was believed to be the female and the smaller (W2) the male. *Photo by Cornell University*

certain W1 was a female and W2 a male since eagle females are larger. It is believed that this sexual dimorphism has evolved in raptors because it is the female who tends to sit on the nest for long periods of time and defend the young against predators, while the male is the primary hunter who depends on dexterity more than size.

Like any parent, I was ambivalent about the eaglets' impending independence. While excited to see them try their wings, experience freedom, and return to their rightful place in the sky, I was nervous about their ability to be on their own and avoid all the threats that could befall them. I had protected them to this point, but I knew it was time to let go. Their flight feathers were fully developed now, their wing muscles strong enough to lift them more than a foot above the nest. Their desire to escape through the bars by launching themselves at them had resulted in frequent crashes, which portended possible injury.

The day finally came to let them test their skills. First, we attached a branch to the platform so it extended out a few feet; this would function as a practice perch. Then we removed the bars. In the wild, young eagles often leave the nest cavity and spend time on branches of the tree, exercising their wings and leaping in the air, before they take their first flight. By afternoon, one of the birds was out on the perch, flapping his wings and jumping in place. So far, everything was going well. Tall oaks and maples lined the edge of the woods about seventy-five meters from the platform. I prayed their first flight would be to one of these trees or, even better, to a dead snag near the tower where they wouldn't have to contend with leaf-laden branches.

I should have known better. Neither the birds' confidence nor coordination had developed enough to land on a perch. Instead W2, the first to fly, headed for the trees, but failing his landing, flew straight down into the open marsh that lay far below the hill, virtually inaccessible from where I was. At first, my heart lifted at the sight of him taking off and gliding through the air, then sank fast as I watched where he was landing. I had no time to celebrate his crowning achievement, to applaud his greatest accomplishment.

It made sense that the marsh would become my nemesis: an easy place to land with no obstacles like trees to contend with, dotted with small pools hidden between clumps of thick cattails, allowing no air to penetrate. Jim and I had put the radios on the birds just in time.

Their maiden flights were a bit like watching a baby's first steps or a child's first bike ride—lots of flapping of limbs, nothing graceful, and much left to chance. A young eagle's wingspan works both for and against a perfect landing. He doesn't exactly crash since his flight feathers provide enough lift for him to glide softly to the ground or a tree limb, but neither does he have any control since his wing muscles aren't developed enough. Once the eaglet settled into the cattails, he was there to stay. The grasses were too thick for him to move through, and his feathers too wet from the pools scattered amid the clumps of vegetation.

I hadn't anticipated this disaster—*but how could I not have foreseen it?* I hopped into the Scout with my transmitter and its cumbersome antenna that resembled the bunny ears of a 1950s TV. Racing along the dirt roads, I tried to drive as close to his landing spot as possible. I had no GPS or map but even if I had, these farm lanes wouldn't have been on them. I depended instead on the beeps of the transmitter to guide me. When the signal was strongest, I stopped the Scout, pulled on my hip boots, and headed into the cattails, holding the antenna and transmitter in front of me. The beeps grew louder the closer I came to the bird even though I couldn't spot him through the curtain of reeds. I followed the sound, stumbling over the hummocks of marsh grass and frequently collapsing into invisible deep holes.

The day was hot, well into the 80s, with high humidity, and no breeze to provide relief. Mosquitoes were feasting on every exposed part of me, delighted with my sweat. I regretted my decision to wear a short-sleeved shirt that morning, far better for heat than bugs. My hip boots filled with water with each fall, making progress even harder. My legs felt leaden, forcing me to tread in slow motion through the cattails. I might have given up had not the adrenal rush from the fear of losing one of the eagles driven me forward. *How could I account for such a tragedy to everyone invested in this bird's future?*

Finally, I saw him, looking as disheveled as I was, squatting in an opening. His eyes reflected a mixture of terror and helplessness. In my backpack, I carried a large cloth, the blue-jean sleeve, and electrical tape.

He tried to fly but gave up quickly and flopped back down into the water. I knew there was no hurry now, so I moved in slow motion to avoid scaring him. After throwing the cloth over him, I wrapped my arms around him in a tight hug, keeping his talons aimed away from my body. Groping under the cloth, I grabbed his legs and held them together in my right hand, using my left to entwine his feet with the tape. Now I could ease him into the sleeve while removing the cloth at the same time. I was sure he could feel my heart pounding throughout this process, but once I felt I had him under control, my muscles relaxed and my breath began to steady. Securing him under one arm while carrying the antenna in the other, I navigated my way back to the Scout, stumbling a few times over the hummocks but somehow managing not to fall and drop my treasured parcel.

I had learned during my zookeeper days that animals will pick up on your emotions, especially when they are afraid. Even zoo animals known for their fierce temperaments will relax if their captors act calm and nonthreatening. This knowledge helped me temper my feelings as W2 and I navigated our way back to the Scout. Not that in this situation it did any good.

On our drive up to the nest site, the fledgling let me know what he thought of our adventure. Upset beyond measure, he clacked his beak at me repeatedly, mixing in a guttural hiss now and again during the half-mile journey. I had saved him from danger but wasn't going to receive any gratitude. I wanted to explain to him that even though, like his parents, I could have fed him on the ground, unlike them I couldn't protect him from predators. A flightless young eagle would have made a tasty meal for one of the water snakes, snapping turtles, mink, and other hungry creatures sharing the marsh with him.

Once back in the nest, the escapee spent some time preening his feathers to set everything straight after his rough journey, but within the hour he was back out on the perch. With any luck, he had had enough excitement for one day and would wait until tomorrow for another flight attempt. Meanwhile, his nest mate W1 watched passively, waiting her turn for a great escape.

## — Chapter 14 —

# A Sudden Goodbye

Our marsh adventures went on like this for several days, always ending with a search and rescue operation and return to the tower. It was a relief when the eagles eventually decided to head toward the trees nearby and try their luck at a normal landing. Their first couple of attempts were mini disasters, sometimes missing the intended perch and dropping several branches down. In a few days, though, they were able to fly from tree to tree and finally figured out how to make it back to the tower by themselves. I continued to put food in the nest to ensure their loyalty to the site.

I fed them on the platform for several more weeks through the month of August. When they weren't in the nest or in the trees around the tower, they explored the marsh, trying their best to snag carp on their own. I tracked them wherever they were and observed them through a spotting scope, recording all their behaviors. I was desperate to see them catch a fish to assure myself they could be independent.

Much of my research time had now shifted from the hilltop to the dike road bordering the Storage Pool below Clark's Ridge, which consisted of a cattail marsh, dead snags, and open water. Not wanting to leave the dogs alone on the hillside for these long periods, I loaded them into the Scout and took them with me. As I observed the birds, Max and Moses slept in the shade of the vehicle or nosed around the shallows looking for critters. When it was near dusk, muskrats and raccoons provided them with entertainment as the mammals left their hiding places to hunt for prey, leaving tracks in the mud along the edge. The dogs were oblivious to the eagles and vice versa. They seemed to have a mutual understanding and respect for one another.

For hours at a time, I focused my scope on the birds as they hopped through the cattails or rested on the dead snags dotting the marsh. In the heat of the day, the young eagles napped frequently, sitting motionless on their perches, just as they had in the nest on the hilltop. My attention often drifted when things were slow for long stretches, and I trained my binoculars on other birds fishing the pool, like Great Blue Herons or mergansers. This always ended with me berating myself when an eagle flew off, and I hadn't seen it depart. It took time before I could find it again, and I worried that I might have missed something significant. I forced myself to keep them in sight; my vigilance finally paid off.

One day, as I scanned the marsh and snags for the two eagles, W2 appeared in my binoculars walking about in the shallows. His random amblings, so haphazard up to now, appeared to be purposeful this time, so I watched the water around him. The back of a fish broke the surface of the pool, dove under, and then came up again in front of the bird. Sure enough, W2 was stalking a carp in the shallows. Suddenly, he was lunging at it with talons bared, digging into its scales until he had a good grip, then hauling it up on the beach where his beak and talons finished the job.

One of my charges had just made his first kill! Pride, relief, and pure joy coursed through me at that point, as my greatest concern—that young eagles had to be taught by their parents to find and kill food—was allayed. Even though I dreaded the time when the fledglings, now independent hunters, would leave the nest site altogether, I knew at that moment we had succeeded in raising these birds to become self-sufficient, facilitating their chances of survival.

---

By mid-September, I was continuing to leave fish in the nest and climbed the tower each day to see if the birds had been there. Their transition from dependence on me as surrogate caregiver to total independence happened quickly. Once they had learned to catch fish, they had no need or desire to return to their nest for food. I wondered if wild eaglets abandoned home so suddenly. Was this normal behavior for Bald Eagles or the result of being raised without parents? Whatever the answer, the only factor that seemed important was their survival. So long as I could observe them in the marsh, I knew they were safe, and I could rest easy. For a week or so, they let me enjoy my fantasy.

Some of my attention during this time was diverted by the unexpected arrival of a third eagle from the Midwest in early September. A two-year old female, who had been rehabilitated at the University of Minnesota Veterinary College, was sent to Montezuma to be released with the hacked birds. Upon arriving in Syracuse, the Peregrine team brought her to the Hawk Barn in Ithaca where they outfitted her with a USGS band and color wing tag. Once we deemed her healthy and ready for the wild, we released her in the Storage Pool where she spent a few days with W1 and W2. The three birds seemed to be getting along well until one day the older eagle unexpectedly left the refuge. She wasn't seen again until three weeks later when, tragically, her body was recovered tangled in barbed wire at a farm several miles away. She was the victim of one of the many hazards that can befall raptors who hunt prey along the edges of fields separated by barbed wire fences.

---

What I hadn't counted on was the Wisconsin eagles being harder to track and monitor once they stopped coming back to the platform for food. One morning I drove to the dike road next to the Storage Pool where W2 had made his first kill. No signal was coming up on the radio receiver, and no eagles were on the snags, their usual morning perches. I drove up and down the road with my antenna out the window, desperate to hear a beep. When no sound was emitted, I searched the roads outside the refuge, thinking that they might have flown further now that they were getting stronger. A clear beep finally arose several miles outside the refuge boundary, in a stand of trees bordering a plowed field.

Over the next several weeks, I followed both birds in the Scout as they moved back and forth between the farmland, the refuge pools, and the forest that lay in between, an area of several square miles. My days were spent driving the roads searching for a signal. The beep frequency for each eagle was slightly different, so I had to figure out which bird I was hearing and if only one, continue driving to try to pick up the other one nearby. My worry about their welfare mounted whenever I couldn't locate both eagles. During this period, I was a wreck, always imagining I had lost them until, after hours of driving in circles, I would finally hear a signal. Usually they were close together, neither wanting to be far from the other.

Autumn soon came to the refuge, and I wondered how long the eagles would stay close by. There are so many unknowns when it comes to Bald Eagles and migration, with most of the available data depending on banding recoveries of birds found dead in states far from their nesting territories. In the case of hacked birds, there was no data, so all I could do was guess. Would they migrate south in search of open water once the pools and streams froze? Would they stay in the general vicinity but move around in their search for food? I could try to locate them every day but foresaw the time was coming when I would no longer be able to ensure their safety.

Hunting season in the refuge and the surrounding area was scheduled to begin October 1st, and as this date approached, something had to be done. We had no idea how the eagles would react to seeing hunters in their marsh when the pair had become used to having the area to themselves for three months. Although I didn't worry a hunter would knowingly shoot an eagle, my concern he might mistake an eagle for a duck, although unlikely, was still there.

With Tom Cade's and DEC's help, we convinced the Montezuma refuge manager, Sam Waldstein, to postpone the start of waterfowl hunting season until mid-October, two weeks beyond the state's normal opening day. This was a big decision for the refuge since hunting season dates were posted a year in advance, and hunters purchased their licenses based on those dates. However, the Montezuma staff wanted the eagles to succeed as much as we did and so were understanding of the situation. They agreed to postpone the start until October 14th.

I still worried duck hunters in the marsh area would scare the birds into leaving before they were ready. Sure enough, on the first day of hunting, W1 and W2, who had been hanging in and around the refuge all fall and fishing in the Storage Pool, took off a few hours after the hunters' arrival that morning. I will never know if they were driven away by the gunshots or the presence of hunters near the pool. Their departure had a note of finality to it, but I made valiant attempts to scour the countryside for radio beeps. On that day, however, they had moved out of the area completely, and I could no longer pick up their signals. I didn't see them again after that. I took solace in knowing they were feeding themselves but was still bereft at losing contact with them. To everyone else that fall, they were a success story; but to me, as much as I hated to admit it, they still felt like my birds.

I am not a hunter, and although I acknowledge the contributions (through license fees and duck stamps as well as the work of Ducks Unlimited) of waterfowl hunters to habitat protection, especially the preservation of wetlands, I was distressed that the eagles had been scared off in this way. I knew that it was a matter of time before they would have left anyway, but it still rankled me that our connection was severed so abruptly.

I remained on the hillside for another two weeks on the off chance they would return, but they never did. Their sudden departure was hard for me to accept. My mind became obsessed with images of them hurt or in trouble, of not finding enough food, of being lost and unable to find their way back to the refuge. I fought against these thoughts, knowing they weren't productive for me or the eagles, but I couldn't help being riddled with worry. I had no way of knowing if they were OK since their radios were out of range. I just had to hope that they would survive. My job had been to raise them to independence, and I had done that. Even an eagle parent couldn't guarantee its young would avoid all the perils of life in the wild. We both relied on faith that resilience and strength would be enough.

## — Chapter 15 —

# Eagle Politics

Unlike the Bald Eagle, one of the few raptors whose range is strictly limited to North America, the Peregrine Falcon is found in many parts of the world. Falconry has been called the "sport of kings" because of its origins with nobility who used various birds of prey for hunting. Today it is practiced by people from areas as diverse as the United States, Europe, South America, the Middle East, and Mongolia. Although Peregrines are not the only falcon of choice, their speed and agility make them one of the most coveted and respected.

By 1976, the Peregrine Fund's success had captured the notice of not only Americans but also European and Middle Eastern falconers, who shared a love for the species. The project was captivating because this was science the public could understand. There were no hypotheses, labs, test tubes or microscopes. The Peregrines were a success story built on one man's dream and the skills of his team, who worked behind the scenes to make it happen. The falcons' later visibility in places like New York, Boston, and Chicago didn't hurt their public relations. No longer a target of conservation, the birds had become the embodiment of its possibilities.

It was not surprising, therefore, that the US Fish and Wildlife Service placed the initial responsibility for the Bald Eagle restoration program in the hands of Cornell's Peregrine Fund staff. Although it was agreed that New York State's DEC would eventually take it over once the logistics and inevitable wrinkles had been ironed out, Cornell biologists were to be in charge at the start. It was to be our design and operation for the first two years. By the end of 1977, in my role as the on-site graduate student researcher, I was to have put in two field seasons and would then be off to write my thesis about Bald Eagle hacking. My job was to start the program and then hand it over to the state.

So captivated was I by the romantic notion of rescuing this iconic bird, I couldn't foresee that the Peregrine Fund's reputation, combined with the stature of the national symbol itself, would launch the Bald Eagle project into the public spotlight. I buried beneath my consciousness all potential hurdles. This was my survival ploy, for had I recognized the truth of my situation, I might have never made it to takeoff. Thankfully, all publicity surrounding the reintroduction project was held at bay until October 1976, after the two eaglets from Wisconsin had successfully left the refuge, alive and well.

I held a crushing fear of being on center stage, of being exposed as the introverted person I was. I felt like a poseur, pretending to have everything under control but feeling like I was winging it (no pun intended) every day. Fear that I would lose the birds, either before they even flew from the nest or afterwards when they headed into the marsh below the platform, fed my nightmares as I dreamt about the articles that would be written on the incompetent eagle murderer. And yet, although I was aware that publicity was essential and inevitable to save this bird, I fought against it when it was aimed at me instead of the eagles and the project. I chastised myself for years afterwards for not being able to "get over myself," but I couldn't, not then.

Fortunately, reporters couldn't find me. Although drivers on the New York Thruway could see the nest platform in the distance as they passed, the Clark's Ridge site was in a restricted area, inaccessible to the public and hidden amid a labyrinth of dirt roads and fields within the Montezuma refuge. While this isolation was difficult to deal with on a personal level, it saved me in 1976 by giving me space to succeed or fail without prying eyes dissecting my every move, whether brilliant or idiotic.

By 1977, my honeymoon was over, and the media became relentless. Once they had heard about the success of the previous summer, press releases covering the eagles abounded. After the Wisconsin pair had safely left the area that fall, the existence of the bicentennial rescue mission was announced on the front page of *The New York Times*. The young female graduate student who had cared for the birds while living alone on a remote hilltop with her two dogs became the new story. Requests for interviews surfaced daily, but I could always deflect them by using the birds' safety as my excuse. Reporters had to go through the refuge station to reach me, so I became adept at not being available.

Often, I would refer news people to Tom Cade, whose familiarity with the media had served him well in the Peregrine project. While Jim avoided dealing with politicos, media, or red tape, and wanted to be left alone to focus on the birds, Cade knew that appearances are everything when it comes to funding an operation of this scope. The public was an essential evil that had to be dealt with carefully, like it or not.

He was willing to do radio and TV interviews on the local news, while I dealt with many of the print journalists. But one paper, the *Syracuse Post-Standard*, was more persistent and wanted to do their Sunday magazine story with me on the cover. Could they send their photographer to see me? I did consent to an interview with the article's writer, recognizing that the publicity would help the species' recovery in the long term. But when I refused to have my picture taken and instead suggested they put the eagles on the cover, they laughed and said, "But *you* are the story, not the eagles." I didn't believe they were serious and held my ground. I had worked too hard to defray the importance of my gender in the raptor world to have it used in this way. I knew that if I were a male, I would not be cover material.

I admit that my recalcitrant behavior was true to my personality. At the age of twenty-six, I hadn't yet learned (unlike Cade) to accept that publicity was a necessary aspect of the politics of wildlife conservation. To me, eagles were sublime and majestic, and I couldn't fathom why people couldn't identify with them like they could with those who studied them. Even as I read about Goodall's work with chimps and Dian Fossey's work with gorillas, I didn't fully comprehend the impact their personalities had on their projects' required financial support.

Understandably, the DEC administrators wanted to advertise the program that they were in line to take over. I managed to antagonize my superiors, albeit unintentionally, with my obstinance. After going back and forth with the newspaper concerning the article and continuing to refuse to pose for the cover, I believed the eagles would become their new subjects.

Removed from all media outlets on my remote hillside—a virtue of the pre-computer age—I probably would have never known about the outcome. But I wasn't so lucky. A month later, one of the people working in the refuge office brought me a copy of the paper. Imagine my shock when I saw a female DEC employee on the cover, in living color, the eagles in the background, but my name mentioned throughout the article as the "mother hen" to the eagle babies. After that, I was determined to limit all interviews to scientists or conservation-minded journalists, and to stay away from anything that smacked of "human interest" where I was the human.

Looking back, I realize I should have accepted the role as the eagles' spokesperson and not left them on their own. But, simply put, I wasn't ready. I regret some of the decisions I made back then when I forfeited all chances for fame that were offered. I even turned down TV coverage. The game show "To Tell the Truth" invited me to come on as the lead contestant. I would have had to tell my story, not just about my eagle work but my personal journey leading up to it, alongside two imposters pretending to be me. The celebrity panel would ask each of us questions and then guess which one was telling the truth. If they were stumped, the contestants would receive prize money. In the mid-1970s, it was a hit show, and would have engendered a good deal of publicity for the project and for me. But true to form, I declined the invitation. The image of Bud Collyer, the host, relating my story while I stood like a statue in front of a studio audience was too much to bear. Today, I regret not having the tape of that show to give to my children and grandchildren.

## — Chapter 16 —

# Without a Stetson

By mid-October 1976, with autumn in full sway on the marsh, I returned to my apartment in Ithaca. Although I continued to stop at Montezuma every few days to make sure the eagles hadn't been back to the tower, I began to reacquaint myself with the glorious, mundane aspects of my former life. Seeing friends, eating fresh vegetables, fruit, and meat not from a can, having a daily shower, taking full advantage of indoor plumbing—all were luxuries to me after months making do.

Soon after returning to Ithaca, I received a call from Tom Cade inviting me to his office the following day. I assumed we were debriefing about the summer or reviewing my academic plans for the next school year. With his help, I had to do some fast footwork to design a class schedule that would both allow prep time for the next field season at Montezuma and fulfill my degree requirements.

His wide smile as he tilted his chair onto its back legs might have disarmed me had I any clue what was coming. But instead, I just assumed he was glad to see me and pleased with the project so far. After all, the birds had survived the release and had even been seen hunting before they dispersed, an auspicious sign. The hacking techniques had worked well for the most part and what hadn't, we were able to remedy easily and quickly on site. All the press releases and video footage of the project promoted by New York State were positive, casting a glow on the work being done by Cornell and The Peregrine Fund. Tom Cade had proven himself not only the champion of Peregrines but now of Bald Eagles, at least of two of them. It was a solid beginning.

So, when he pushed the flyer towards me on the desk, I must have looked confused. A photo of a raptor—perhaps a Peregrine or eagle, I don't recall—was on the front with the words "1976 Raptor Research Foundation Conference." Then he put his finger on the line below and tapped it slowly: Ithaca, New York, October 29—November 1. His smile returned. Dr. Cade was not a smiler by trade. His was a serious, almost stern, visage much of the time, so again I was unnerved by his expression. Then he lowered the boom: "I've signed you up to give a presentation to the RRF about eagle hacking. I think it would be a great idea to share our success with everyone and toot your own horn a bit."

My response was swift and uncalculated. "What? I can't do that! I'm terrible at public speaking and besides, what would I say? Don't I need slides? I haven't compiled my data yet! I have no concrete results . . ." I stammered, panic rising like volcanic lava on the brink of eruption, as I groped futilely for more excuses.

Cade remained calm, probably expecting this reaction, and in a soothing voice reassured me: "You know this subject better than anyone, and you have a responsibility to share your knowledge with the raptor community. I know you can do it, or I wouldn't have put you up to it." He wasn't leaving room for more objections. I knew this wasn't a repeat of "To Tell the Truth," when the decision had been up to me. I couldn't say no and risk losing the respect of this man who had given me such an opportunity. I felt bile in my throat that I struggled to ignore as I grasped the reality before me.

---

The Raptor Research Foundation (RRF) was created in 1966 in response to the large-scale poisoning of all populations of birds of prey. Most raptor biologists were members, as well as others who cared deeply about the birds. They met once a year in different locales, usually in the West or Midwest, where researchers presented papers and shared their findings with others. I had never been to a meeting but could imagine the crowd. I knew the majority of the 150 to 200 members attended this conference. If I had had trepidations about walking into the Hawk Barn and dealing with the men-only club of the Peregrine crew, the assemblage at the RRF meeting would be far more frightening. I glanced at the schedule of events in the brochure just long enough to see my name and the title of my presentation listed for Saturday, October 30, the first day of the talks. At least it would be over and done with early. *But I had less than two weeks to prepare! How was I going to create a presentation and pull myself together in such a short time?*

The next fourteen days were a blur as I collected slides, organized them into some type of order, and wrote up my talk. Before computers became available to us after 1977, photos had to be loaded by hand into a slide carousel and then practiced over and over with a projector to be sure none of them was upside down or out of sync. By the time the conference started, I had rehearsed hundreds of times, knowing that my only chance to get through this ordeal without choking was to know my talk cold.

My greatest fear was that the slides would sabotage me, leaving me abandoned at the podium with a blank screen behind me, my words stuck in my throat. Once you were up there in front of an audience, there was no fiddling to make things work. It had to be right from the start. I had been to talks where a slide stuck or was skipped and the entire talk went off kilter, the presenter left to suffer, helpless and alone.

---

A ragged symphony of slamming doors rang out from the Ford pickups and Wranglers crowding the Ramada's parking lot. The hotel's portico loomed ahead, beckoning the arrivals into the Emerald City. Throngs of men—I don't recall seeing any women, but there must have been a few—sauntered towards the entrance, stopping to slap backs, push Stetsons back on their heads, offer hearty handshakes to what seemed to be every person who passed them by. I tried to breathe deeply to steady the nerves jangling throughout my body. To no avail. *I couldn't do this, it was a step too far, how could I have ever thought I could come here?*

I had arrived at the Ramada early that morning, even though my talk wasn't until just before lunch. That was a rookie move, starkly evident when I drove into the parking lot and saw all the glad-handing going on. Besides a few members of the Peregrine crew, I knew nobody in this group, but they all appeared to be old buddies. I had given myself more than enough time to work my brain into a frenzy.

Feeling notably conspicuous as I wandered the hallways of the hotel alone, I became more agitated when I sat in on others' talks, comparing mine, which seemed like "Tales of my Summer Vacation," to their meticulous tables of hard data and graphs of trends. I had pictures, pretty pictures I admit, of the two birds, the marsh area, the tower, and other tangibles. But my research was far from being quantified, and I could provide no statistical analysis for these scientists in the audience. *What would they think of this woman daring to teach them about hacking eagles?* I could visualize the wry grins, raised eyebrows, and nudging of elbows in the seats.

The main auditorium, where all the speeches were presented, was cavernous and filled to capacity. By the time I approached the podium, hooked up my carousel to the projector, and tested the microphone, sweat dripped from every pore. I looked out at the audience of raptor scientists, relieved to find the room dark, their faces obscured. The moderator introduced me with words that could have been in a foreign tongue, for by then my heart was playing the bongos, drowning out all other sounds. Once I saw his lips stop moving and he backed away from the microphone, it was as if someone had pulled a cord from the back of my neck; I began to talk. The slides saved me. Everyone loves looking at eagles, and my topic, as Cade had reminded me, was fresh and compelling. I felt from the beginning that the crowd was engaged, the passion that had invaded me on the hillside all summer reinvigorating me on that stage. My terror subsided, replaced by the realization of the importance of our work and the need to prove to these men that it was possible to save this bird.

As soon as I made it through my final slide—an iconic shot of an adult Bald Eagle, head and tail gleaming white, soaring over open water, reminding everyone of its magnetism—I turned to exit the stage. Relief flooded over me until I heard the moderator's next words: "I think we have questions from the audience. I see a few hands in the air."

*Questions?* It wasn't just that most people in the room were either professors or government biologists specializing in birds of prey. It wasn't just that many of them were falconers with extensive raptor field experience. And it wasn't just that I was being thrown into this esteemed group after a mere four months in the field. It was more that I was being given a final exam after the first week of school.

Fortunately, the questions were not aimed to trip me up but just to find out more about how the eagles reacted to the reintroduction process. Simple questions that I could answer with a modicum of authority: *How did they react to different food, how did I know when the birds were ready to fledge, how well did the transmitters work?* One man who identified himself as an eagle bander from Alaska even asked about the blue-jean sleeve, my greatest invention. He wanted to know what had made me think of it. Laughing, I replied, "I can't sew so I was looking around for something ready-made, the diameter of an eagle, with openings at each end. It turned out I was wearing it!" As the questions went on, I became aware I had the answers, and everyone wanted to learn from me. This revelation calmed my nerves and instilled in me a confidence I didn't know I had. The voices were kind, laced with compliments about the project's success, and more curious than probing. I sensed a collegiality from the group I hadn't anticipated.

The most surprising part of the day was the aftermath of my talk. I was surrounded by men asking more questions or sharing their own knowledge of Bald Eagles. Most were from Alaska or the Midwest, decked out in red and black flannel, some with string ties and silver belt buckles. But there were also a few from the Great Lakes or the South—Florida and the Chesapeake—where Bald Eagles were still plentiful enough to provide viable population and behavior data. They shared their stories about banding eaglets, being dive-bombed by their parents, watching the adults feed the young on the ground, and observing wobbly first flights. Soon we were a close-knit group with a shared passion for Bald Eagles, and finally I began to think of myself as a raptor biologist. I probably learned more about eagles during those few hours after my talk than in all the book research I had done.

And the contacts! I put faces with names that I had only heard or read about—the fathers of eagle biology. I quickly forgot I was the only female in this elite group of eagle scientists. The bird was the catalyst that was throwing us together and that would bind us for the rest of my tenure at Cornell. I felt I had found a home at last among people who shared my love for the work. It was obvious that this was a cohesive group that had met each other before, whether in the field or at other meetings. They formed a fraternity I was desperate to join, but recognized I had years to go to merit admittance.

The conference couldn't have happened at a better time, for my networking that October not only proved essential for the next season's work but also provided the burst of energy I needed to repeat the program in 1977. The fatigue that had descended on me by the end of the first summer dissipated once I could see how significant and groundbreaking our work really was. The excitement evident in the audience's reactions was proof enough that this project transcended the political headlines of saving the national symbol. American scientists from Alaska, the Great Lakes states, Illinois, and North Dakota were mixing with Canadian biologists from Saskatchewan, Manitoba, and British Columbia. To the members of the RRF, eagles were important in their own right, being one of the largest raptors endemic to North America, and one of the most threatened. What we had proven counted as a successful experiment that could be reproduced to ensure the survival of a unique species.

The reverence that these men felt for their bird—matched by the Peregrine scientists and other raptor specialists who dedicated their lives to their species of choice—was both palpable and contagious. All the doubts that had plagued me since I had first been offered the chance to save the Bald Eagle a year ago were gone. My inexperience, gender, and lack of confidence no longer seemed important. I left the conference that day filled with pride and a profound sense of accomplishment, as Tom Cade had predicted.

# Part Three

# 1977

# — Chapter 17 —

# Changes

From my distant vantage point in Ithaca, I could see more clearly the close calls, faulty judgment, and innocence that plagued W1, W2, and me in our common pursuit of survival during that summer of 1976. It was not lost on me that the attention I received at the RRF conference from the eagle biologists—as much as I would like to think it was personal—was due more to their professional curiosity. They displayed palpable relief that Bald Eagles can survive in the wild without their parents, that the normal rites of passage to becoming a self-sufficient adult occur by instinct and don't have to be taught by parent birds. The successful departure and independence of the Wisconsin birds, even though there were only two of them, were visible proof that hacking eagles could be done.

While I had feared being alone for three months, by August I had come to relish my independence. Although I didn't realize it at the time, during that first summer I had been given the gift of autonomy as I studied the birds, explored what worked and what didn't, and had the luxury of correcting my mistakes. This was feeling like *my* program, and the eagles, *my* birds. For the most part, I worked alone even though Jim and Willard were a phone call away. I could easily forget that there were many others invested in the eaglets' future and that I was merely a temporary custodian of the project.

I had proven to myself that I could do this job, which, one year ago, had seemed way beyond my abilities. Although I never spent time celebrating this accomplishment, I was relieved beyond measure I hadn't squandered the opportunity presented to me. In many ways, my two charges in the nest had been guinea pigs as much as I had. The missteps during that first summer were mine to make. I had to gain the confidence of the federal and state agencies who were funding me, even though it meant surviving many wrong turns before I could believe I had done this. Everyone connected to the release was beginning to view the program as less of an experiment and more as a long-term conservation model.

---

The summer of 1977 was to be different from the first summer in every way. By mid-May, we learned that Wisconsin eagles had not produced as many three eagle nests as in the preceding year. Usually this has to do with food supply. Although parent eagles will lay only as many eggs as they think they can support, sometimes they misjudge either the extent of their larder, the weather, or some other unpredictable factor. When this happens, only two of the three nestlings might survive, the third starving as his stronger siblings grab all the food or toss him out of the nest. As was supposed to be the case in 1976—although the banders removed W1 and W2 from the same nest—the 1977 program was based on the premise of rescuing the third eaglet who may not have made it otherwise. Translocating eagles is reliant on politics and perception. It was important to explain to the public that we weren't removing healthy eagles from the Midwest's population but were rather saving eaglets who would not have survived without help.

With Wisconsin out of the running for eaglet donations, the US Fish and Wildlife Service, working closely with Tom Cade and DEC, turned to Michigan and Minnesota, states glad to donate a couple of their birds to the project. These eagles had had a good year, and there was a decent selection of three-bird nests to choose from.

Although there was no established captive breeding program in the works for eagles, as Cornell had for Peregrines, there was a nascent program at Patuxent Wildlife Research Center in Laurel, Maryland. The Bald Eagles there were under the direction of biologist Stanley Wiemeyer. After the 1976 releases proved successful, Cade received a call from Stan, who was interested in using his captive-bred eaglets in the hacking program. In the spring of 1977, I flew to Washington, DC, and drove out to the research center to visit Stan.

# Changes

The trip to Patuxent was life-changing for me. As I flew out of the Syracuse airport and drove a rental car to Laurel, I felt the mantle of professional biologist had finally settled on my shoulders. To spend the day at Patuxent, discussing with Stan the possibilities of using captive bred eaglets for the next summer's project, made the past year of worry, doubt, and isolation worthwhile. I was part of the team entrusted with the future of this program. We toured the state-of-the-art facilities and observed the adult and immature eagles in their spacious flight cages. Stan was a slim, soft-spoken man with blonde hair and wire rimmed glasses, and an enthusiasm that was in sharp contrast to his otherwise calm demeanor. He viewed the hacking concept as justification for breeding Bald Eagles in captivity. Curious as to how his young birds would fare when removed from their parents and released to the wild on their own, he eagerly agreed to supply his first two eaglets to the New York program. This was an act of faith on his part since these would be his first two hatchlings, his breeding adults having had no success in producing young the year before. Meeting Stan and the adult pair of eagles who parented my adopted charges made me feel even more responsible for the success of the project.

We now had the promise of four eaglets available for release: two from Patuxent, one from Michigan, and one from Minnesota. My foster number had doubled and with that, we needed to expand the nest tower. Four birds in one nest would be too many.

In May, we erected a second tower next to the first, built a second nest, and enclosed it with the same barred sides as the adjacent one. My observation blind had to have access to both nests now. By placing it between the two platforms but about two feet lower down, the birds could see each other over its top while I could still feed and watch both sets of eagles. The scaffolding that had supported the blind in 1976, and had provided me with a sturdy structure to climb, had to be removed to make room for the second nest platform. Now I had to climb up to the tower using the steel pole steps sunk into the telephone poles as hand and foot holds—the same as the ones used when we had built the original nest. At this point, I was so used to the climb that this rudimentary system didn't faze me as much as it would have the previous summer. I had become part lineman part mountain goat after so much practice. My fear of heights had evolved into a healthy respect and sense of caution when above ground. But I still never looked down.

As in 1976, we weren't given any warning when the eagles would arrive until the day before and had to rely on the banders to determine when and where the young eaglets would be caught and sent to New York. I had to be prepared to meet the birds at Montezuma as soon as their plane landed in Syracuse.

Normally, this wouldn't have been an issue, but that summer I was supposed to attend my oldest friend's wedding in mid-June on Cape Cod. Although I had told her my availability depended on the eagles' arrival, we had both hoped it would be after the wedding so I could attend both events. I had two sets of bags packed—one for the eagles and one for the wedding. I didn't know until the day before whether I would be back on the hillside in my blind and tent, feeding my new charges a dead carp, or on the Cape teetering around the dance floor in heels. The banders collecting the eagles didn't plan around my calendar, however, sending the eagles to me on June 24, the Friday before the wedding.

---

The DEC administrators, expecting to take over the program after the 1977 season, began to become more engaged in the process. This made sense from their point of view, but their heightened level of involvement caught me off guard. I knew the increased number of eagles would pose challenges for me in 1977, but I was unprepared for some of the other changes awaiting me.

First, my employment had been switched from Cornell to DEC, so I was now a state employee and no longer an independent contractor. I didn't understand until later what that change of status would mean.

Earlier in the spring, DEC told me I would be living in a small trailer instead of my tent. It would have a kitchenette, a bunk, a table, and solid roof. No more leaky canvas, Coleman camping stove, or sleeping on the ground. At first, I reveled in the idea of such luxury, for as inured as I had become to rough camping, the idea of a bunk was appealing. My own little house that didn't leak and rock in the wind.

The next change to my world involved the video camera. The photographers, Lance and Pat, were rehired in 1977. They were now contracted to install two video cameras on a pole next to the tower, each focused on a nest so the eagles could be filmed remotely 24 hours a day. This would feed into a monitor in the trailer. The purpose was two-fold. I could watch the birds continually from the comfort of my new abode, and it would also provide a tape of all their activities so anyone else could have access as well. Again, I thought this was a good thing, at first. I wouldn't have to sit for long hours in the blind, hunched over and cross-legged, recording in my notebook every move they made. Instead, I could climb the tower only to feed them and then descend to the comfort of my trailer to watch them from my new "office".

I hadn't known at the time that future field research would be heading in this direction, that watching animals from a blind was going to become an obsolete methodology. Audiences today can tune into live video coverage of nests on the other side of the world, a concept unimaginable in the 1970s. I had been lucky to experience an old-fashioned but more realistic way of studying animals that first year.

When you are observing behavior up close, there is a connection between you and your subject. The hours I spent in the blind watching the eagles, and listening to every sound they uttered, provided me with a unique perspective on their lives. Although there was a canvas wall between us, I could smell, hear, and see the birds. They may not have known I was beside them, but I knew we were as close to being a family as two birds and a human could be. Although concerns about imprinting kept us physically apart, from my point of view, there was an invisible tether connecting us.

Looking at animals on a video screen separates a researcher and her subjects beyond the physical distance. Sounds are muted, with only the loudest and least subtle vocalizations audible, and it is difficult to tell which bird is communicating. This is made even harder because the eagles can't be distinguished from one another unless their color tags are facing the camera. The odor of rotting fish and bird guano is removed—some might say this isn't a bad thing! Slight movements can go unseen, so some of the communication between the birds is lost. In the case of the eagles, my visual experience was like switching from a TV show in living color to one in black and white. Much was lost in the transition, making my second field season feel very different.

I found out later that one reason for the trailer and the video cameras was that I was no longer going to be operating alone. DEC needed to train its personnel to hack eagles so they could take over the program in 1978 and beyond. To do this, their staff had to learn the procedure by shadowing me for the summer. My solitude was over. Instead, I was to share the 1977 birds with an array of different state workers, the photographers, and more frequent visitors from Albany.

The most regular observers at the hacking site were Peter Nye and Mike Allen, DEC employees in their mid-twenties, who had been selected to take over the Bald Eagle restoration project after I left. Pete worked at DEC's newly created Endangered Species Unit housed in Delmar, outside of Albany, while Mike had been assigned to the DEC's Region 8 office in Livingston County, not far from Montezuma. The two would alternate their time at the eagle site so they could eventually step in the following year. It was ironic that I was supposed to be teaching them even though I only had one more year of experience than they did.

Other DEC employees were also sent to Montezuma to experience the hacking program, but usually only one or two were there at a time. Lance and Pat, the video team, were often at the site, setting up equipment or recording footage for a future documentary on the project. On some evenings, there might be four or five of us around the camp stove, as opposed to the previous summer when it was just me and my two dogs. I enjoyed the camaraderie during those times, connecting with others who felt as strongly about the eagles as I did. We were all about the same age and just starting our careers, a commonality that bonded us. It was only in the light of morning that I found I missed my old solitary routine.

Life had taken a different turn. A possessiveness clouded my role as "hostess of the hillside." After spending a field season with the birds and figuring out how to readjust after each wrong move, it was hard for me to share this with others who hadn't been as involved from the start. Although all the state employees were enthusiastic and helpful, my new role as mentor to the revolving band of observers took some time getting used to. I missed the freedom of waking every morning to watch the sun rise, knowing that the day was mine to plan, the eagles mine to care for, and the eagles' future mine to ensure. I admit I was wary of the influx of people onto what I had begun to think of as my territory even though I had no reason to claim ownership or resent their presence. I had to remind myself continually that the recovery of the species depended on my ability to prepare the next group to take over, that this wasn't a one-person show, and, for my part, a little less ego was in order.

# — Chapter 18 —

# Escape Artists

The high-pitched screams reminded me more of irate monkeys at the zoo than eagles. Then came the gull-like alarm cry I recognized as their "peal call," the sound they make when they feel threatened. The rumbling of the Fish and Wildlife vehicle driving up the road couldn't compete with the eagles' own melodramatic overture.

On June 24, the Patuxent pair arrived at their new home on Clark's Ridge. From the start, I knew these eaglets would be a challenge. Recalling the arrival of W1 and W2 a year earlier, the cacophony coming from the crates today eclipsed the mere wing-banging of the Wisconsin birds. The screams and cries kept time with the cadence of thuds as these two newcomers banged their bodies as well as their wings against the sides of the crates. The wood threatened to split open at any moment, exposing an eagle's wing to the outside.

All of us were aware of the necessity of haste. The sooner these eagles were taken from the crates the better, before they hurt themselves. I did wonder, as the vibrations emanated from inside, how this behavior would play out once they were released. As he had the previous year, Willard tied ropes around the crates. Then, as the clamor increased, Jim hauled the eagles up to the platform. Cade and I were ready with our crowbars to release the young birds from the confines of their boxes.

As soon as the two eagles were freed, they launched themselves at the bars, struggling to wedge their bodies between them. I was astounded how much bigger and stronger they were than W1 and W2. Although both were nine weeks old, the same age the 1976 birds had been, the female was particularly large and completely feathered out. Known as P1, it was evident from the start that she was too energetic and precocious for any life of captivity. The birds were siblings, both born at the Patuxent breeding facility from a father from Alaska and a mother from Wisconsin. Due to their sex dimorphism, there was a big size difference, with P1 significantly bigger than her brother P2. She was broader in chest and taller, making her appear more imposing.

From the start, both eagles were less than happy to be in their new nest and did everything they could to execute an escape, from banging against the bars, to trying to squeeze through the openings, to launching themselves at their caregiver with wings flapping and talons at the ready. Although this was more of a feint, it was still unnerving and made me glad to retreat from the enclosure. It was obvious that the less they saw of me the better since even hearing me move around in the blind sent them into a panic. I was thankful for the video cameras now, for it meant I could watch them remotely and didn't have to disturb them except during feeding.

---

Four days later, two more eaglets joined P1 and P2 when a plane from the Midwest carrying one eaglet from Michigan (M1) and one from Minnesota (M2) landed in Syracuse. Both had been taken from three-bird nests and were examined by Dr. Pat Redig at the University of Minnesota (UMN) Veterinary College. Dr. Redig was an avid falconer whose dual specialty in avian physiology and birds of prey led him to found and direct The Raptor Center at UMN, in addition to being a professor at the vet college. By coincidence, he had attended my talk at the RRF meeting six months earlier and had approached me afterwards to congratulate me. When I heard he had looked over W1 and W2 before allowing them to be shipped to New York in 1976, I felt reassured they had been in good hands. He had also rehabbed the two-year-old eagle he sent to us last September, who had encountered the lethal barbed wire. Now that I was more aware what challenges lay ahead for these 1977 birds, I knew they needed to be strong to succeed at Montezuma.

There was a marked difference between the wild midwestern eaglets and the Patuxent captive-bred birds, not only because of the age differences. Captive-bred birds receive an unlimited food supply, so they can grow faster and feather out sooner. In the wild, the parents of a three-bird nest are struggling to feed all their young, which often results in slower development of the chicks. I wondered often if this different upbringing would affect the eaglets' success in the long-term, and whether being raised in captivity would give birds any advantage.

Pat Redig had another job to do after examining M1 and M2, the two eaglets selected for transport. In one of the nests inspected in Minnesota, the climbers who had captured M2 discovered a third eaglet with a broken leg. Since the injured bird had no chance of survival if he remained in the nest, the climbers removed him and left him with Dr. Redig for treatment and rehabilitation. The hope was that the bird could be released if his leg healed and became strong enough. To make it to Clark's Ridge to join the others, he would have to defeat the odds plaguing wild nestlings with injuries.

From the moment they hopped out of their transport crates, M1's and M2's demeanors reassured me they were far calmer than the Patuxent pair. M1 at eight weeks and M2 at seven weeks seemed more ready to acclimate to their new home. Content to sit in the nest, eat their food, and not try to escape, they were more like the 1976 birds but younger and less developed. I realized there was an optimal age for hacking eagles. They needed to be feathered out enough to withstand the cold nights alone in a nest without parents, but not so well feathered out that they felt ready to fly. The Patuxent birds, more ready for flight and independence, would be teaching us a valuable lesson about whether their age would be a detriment to their success as hacked birds.

---

From the start, P1 and P2 commanded my attention. Too mature to be contained and doing all they could to rebel against their situation, they required some creative maneuvering on my part to keep them calm. My movements near their nest had to be kept to the bare minimum, for any noise I made or, worse yet, the sight of me approaching the tower activated their escape routine. They could hurt their feathers as they launched themselves against the bars.

I assumed they would settle down in a day or so, but instead they refused to eat and focused all their energy on escape. It was three days before they decided to touch their food and even then, they seemed more focused on what was happening outside their cage than on the delicious fish inside. Jim, Willard, and I hunkered down in front of the video monitor watching the two birds thrash about from one end of the enclosure to the other, looking in every direction except at their food inside the nest. "What do you guys think?" I asked the two men. "I feel like they might hurt themselves if we keep them inside, but something else might hurt them if we let them out." I was desperate to have a second opinion for this decision, which could go wrong either way. The three of us considered the pros and cons involved. Both options had risks. If we released them early, they would be unprotected from predators and could risk injury trying to fly before they were ready. If we held them inside against their will, they could also injure themselves, but at least they wouldn't be vulnerable to attack. We opted for keeping them confined.

To minimize their agitation, I tried to make life as easy for them as possible. I moved my trailer further away from the tower so they couldn't see or hear my movements. The increased number of people on the ridge created more noise as well, so the DEC folks restricted themselves to the area around the trailer. I kept the dogs from exploring too close to the nest location because the sight of them threw the birds into turmoil. By shifting my observations from the blind to the video monitor inside the trailer, I had to climb the tower only twice a day to feed them.

The limitations on what I could see on the monitor made my research more difficult. Besides the issues with identifying vocalizations and behaviors of each bird, I couldn't keep track of their food preferences as easily; the items were often hidden below the nest rim. My view was restricted to one nest at a time. Luckily there were two cameras, so when the second pair of eagles arrived, I alternated between nests. We had installed microphones on both platforms, so if something started happening in the other nest, I would know it and switch cameras.

On the day M1 and M2 arrived, P2 became frantic. For four days, I had been trying to keep the disturbance to his world at a minimum, but once the midwestern birds were delivered, it was impossible. There was activity as we raised the new crates to the platform, and much chatter below since the Fish and Wildlife, Cornell, and DEC folks had reassembled to welcome the new birds. This was more than P2 could take. Being smaller than P1 but more feathered out, he was even more desperate for freedom. After thrashing around his enclosure for several hours until dusk, including banging his head against the plywood roof that covered it, he did the unthinkable. Raising his wings over his head to make his profile as narrow as possible, P2 slipped between the bars of the cage. I was stunned at his cunning and credited him for his resourcefulness even though it made my life more difficult. Now the quandary was whether to leave him be or catch him and return him to the enclosure. If I had had the time to radio tag the birds, it would have been a simple decision. I would have left him alone and tracked him. But I hadn't anticipated his escaping so soon. To ensure his safety, he needed to be caught. Because it was almost dark by then, I had to wait until dawn.

As if he knew of my plan, he fell off the tower before daybreak, flapping over to a nearby tree, then flew on to the cattails below the ridge. Out came the blue-jean sleeve! Since he had no radio and the pre-dawn light limited my visibility, it took more than an hour to spot him in the marsh; fortunately for both of us, he hadn't gone far. I drove the Scout down to the dike road adjacent to the spot where I had seen him disappear. There was a section where the cattails grew in clumps, leaving open shallow pools between their stands. The 1976 eagles had usually headed for this spot, most likely attracted to the open water as they came in for a landing. I pointed my binoculars towards these openings, hoping P2 would have followed suit.

Sure enough, his chocolate brown shape stood out despite the thick vegetative cover. Approaching quietly, I tried not to alarm him and risk driving him out of the small pool into the cattails. Despite his feisty nature, he was easy to catch, going limp as soon as I threw a cloth over his head and placed him inside the sleeve. I drove him back to the tower where Jim was waiting. After returning to Ithaca the night before, he had expected to come back in the morning to help me catch and tag the rebellious eagle. Little did he know P2 had made other plans.

We wasted no time in putting radios and color tags on both birds. By now we had a more synchronized system, and I had become less of an apprentice and more of a partner in the banding process. The procedure resembled an operating room scenario: one of us would hold a foot steady while the other snapped on the aluminum band, or one of us would spread out one wing and the other would fasten the color tag around the joint closest to the eagle's body. The tag was outfitted with a radio transmitter encased inside a small pouch we sewed onto its underside. We worked quickly but carefully, knowing that the birds' patience would be short-lived.

After we were done, we left P2 sitting on the platform lip outside the bars so he wouldn't try his squeeze play again, which left P1 still inside, and upset! P2 stayed where he was for four days, eating fish I laid out for him, before he flew again, apparently taking a break before his next adventure. This time he was stronger and went back and forth from the trees to the tower at will.

Meanwhile P1 was working herself up so much that we had to remove the bars for her too. Naturally, she flew towards the cattails when I approached to feed her. Heading back into the marsh, I retrieved her and returned her to the tower. She flew again the next day, but this time she made it back by herself. This recalcitrant pair was finally becoming independent flyers, a huge relief indeed. Each time an eagle disappeared below the ridge, I feared it would be the last time I would see it. Every rescue felt like a miracle, a reprieve allowing me to continue this project. These early days proved to be a test of my staying power. No sooner would I think I had the Patuxent pair settled into a routine when one of them would try a new risky maneuver.

Shortly after both P1 and P2 left their nest, raccoons smelled the carp I had left there. Somehow the predator guards weren't enough to stop the scavengers' quest for an easy meal. They didn't bother M1 and M2, who were still safe in their enclosure, but their presence was enough to prevent P1 and P2 from returning to the tower. I considered it a positive sign that the young eagles recognized the folly of facing off against adult raccoons over food.

## — Chapter 19 —

# Five's a Crowd

Just as things were settling down for me and my four feathered friends on the hilltop, there was an unexpected but welcome development. More than three weeks after M1 and M2 had arrived from the Midwest, the eagle that had been found in the Minnesota nest with a broken leg was pronounced healed and ready for release. Pat Redig and Tom Cade arranged to have the bird flown from St. Paul to Syracuse and driven down to Montezuma to join the others on Clark's Ridge. I was informed of his imminent arrival the day before, on July 21.

M3, as I called him, was eleven weeks old by then and appeared fully recovered from his leg injury. Although he was a week older than M2, he was smaller, leading us to assume he was also a male. Because of his stint in rehab, I was worried he would be strong and ready to fly, and so would react to captivity the same way the Patuxent birds had. Instead of being obstreperous and driven to escape, however, M3 was docile and adapted quickly to his life in the enclosure. Compared to P1 and P2, he behaved almost like a tame bird. There were probably two reasons for this difference.

First, he had been around people at the vet college while being rehabilitated. He had received physical therapy, once his bones had healed, and was exercised by his handlers to ensure his leg would be strong enough to support landing and takeoff. Noises didn't bother him and seeing me near his nest tower had no effect.

Second, he was small and not ready to fly yet. Like the other two midwestern birds, he had been the third bird in a nest and probably hadn't received as much food as the other two nestlings. He was bound to grow slower and feather out later. Both these factors made him the most compliant, and to me, the most endearing, of all the birds in my care.

We were nervous about the reaction of the other eagles to a new arrival so introduced M3 with some reservations. M1 and M2 had been together for three weeks by then. How would they react to a strange bird joining them? In my mind I often compared the eaglets to adolescents, so similar were they in figuring out their world. If this comparison held true, the new eagle might face some resentment.

To reduce the competition and to make more room in the nest, I removed the larger M1, who was showing signs of being ready to fly, from the enclosure and placed her on the platform lip outside the bars. Then I put M3 inside with M2, who was smaller and calmer. The introduction went as expected with a fair amount of flapping wings, clacking beaks, and then passive acceptance. I breathed easier when I could see the new bird was not going to be attacked by M2. By the end of the day, the two seemed at peace with one another, their relationship proving to be far less complicated than my imagined adolescents.

Although not as placid as M3, M2 appeared to be more of a homebody than any of the others, or so I thought. M1, on the other hand, apparently viewed the new eaglet as an excuse to fly the next day to a tree, returning to the tower five days later after spending her time in the woods nearby.

The acrobatics of M2 and M3 gradually increased, signaling when the time had come to release them so they could join the other eagles. By now it was the end of July. The timing would be good because there was usually at least one DEC employee on the site at that point. With two birds being freed at once, I needed another set of eyes to see where they were going.

M3 flew first, landing in a tree near the tower, but then ended up on the ground, unable to become elevated. His complacency came through again, for instead of panicking and running through the undergrowth, he stayed still as if he was waiting for someone to bail him out. He may have been hoping for one of his parents, who would have fed him on the ground in the wild, but he got me instead. I caught him easily, not even needing the blue-jean sleeve, and put him back on the tower to join M2, where he remained for a few more days.

M2, who seemed more attached to the nest site, was taking his time leaving. I hoped that this meant he would be more adept at flight and that my marsh trekking days were over. But no such luck. Two days after M3 joined him back on the tower, M2 made his maiden flight, also landing in a nearby tree but then heading straight for the cattails.

I retrieved him twice more from the same area in the Storage Pool and returned him to the tower. On the third attempt, he managed to extricate himself from the dense cattails and become airborne, landing on a perch in a tree adjacent to the structure. Reluctant to return to the platform on his own, he remained in the woods nearby. He flew from tree to tree, perching for long periods within view of the nest and the food inside it. I believe the presence of all the other eagles—for by now, the four other birds (P1, P2, M1, and M3) were coming and going frequently—made him nervous. It was eleven days before I saw M2 come back to the platform to feed. Perhaps he was scavenging food items on the ground during that time, but his ravenous appetite when he finally returned indicated otherwise.

---

The bond between the 1976 birds was much stronger than with the 1977 birds. W1 and W2 had used the tower as a rec center for weeks after they had become adept flyers. Even when in the marsh, they were seldom seen apart and were always within visual contact of one another. Although W1 remained dominant, as they grew older, they seemed to bicker less over food and were more at ease with each other. This early bond would prove to be a harbinger of their future.

The five eagles in 1977, on the other hand, didn't share this camaraderie. P1 was the most aggressive, and the others knew it. When she was on the tower, the smaller birds were reluctant to return for food. A crowd of five on the platform also might have been a deterrent once each became a skilled flyer. In the wild, an eaglet has only one or two siblings to contend with when returning to the nest, so this was a different scenario.

During the following two weeks, the birds returned to the platform to feed but spent most of their time on the dead snags in the marsh. I never saw them catch a fish, but it was obvious that they were learning about their prey by studying the carp swimming in the water. Ever hopeful, I rejoiced when I saw them leap onto logs as if they were fish, a sign they were practicing their skills. I was a proud mother. The difference was I couldn't help them, couldn't take any credit for their feats and new abilities. All I could do was stand by and watch them take these first steps into independence, steps that would dictate their chances of survival.

## — Chapter 20 —

# Return Home

My fishing skills had to be honed considerably to feed five Bald Eagles instead of two, especially due to the uncooperative weather we had that second season. The summer of 1977 followed an unusually dry spring, lowering the water level in the marsh. The carp, so plentiful and easy for me to net the year before, were now scarce. I had to build a holding pen into which generous fishermen threw carp and other fish they didn't want so I could have enough food for the eagles. I became so desperate that I began to drive the dike road to search for fishermen returning with a catch. It was during one of these encounters I became aware of the secrecy surrounding the eagle releases. A rowboat was heading for shore one day when I was inspecting the holding pen, finding it empty. I waved at the fisherman as he was climbing out, lugging a pail filled with his catch.

"Hi! Have any luck out there?" I asked hopefully.

"Yeah, a few pickerel and bass, but the water's too low for a good haul."

"Did you see any carp? I really need some for my eagles. They're used to having a fish a day, but lately they've been hard to get."

"Eagles?"

"You know. The release project up on Clark's Ridge. I'm the researcher taking care of them, and this year I have five to feed."

"I didn't know about no project up there. I saw the tower but didn't know what it was for. None of us did. I met a couple of my buddies out here the other day, and we were all sayin' we couldn't figure out what was on top of that hill."

"Wow, I had no idea you hadn't been told. It's a reintroduction program for Bald Eagles. I'm raising five young birds and releasing them when they're ready to fly. I know the manager asked some fishermen to put carp in the pen over there, but I still don't have enough because of low water."

"That's some project! And you're doin' all this? Well, I can tell my buddies to look for carp and pop them in that pen if ya like. They're there all right. Those buggers don't mind the low water as much as the natives. Seems nothin' can hurt carp."

From then on, I could count on many of the local fishermen helping me out, supplying the pen with carp whenever they could catch them. I had known the secrecy surrounding the releases that first year had kept the local people in the dark about what was going on. But I assumed by the second summer, everyone must be aware of the eagles' presence.

I also had to supplement the eagles' diet with other animals since depending on roadkill for five birds was too unpredictable. The American Kennel Club donated ring-necked pheasants and mallard ducks they used for their hunting dog field trials held in the refuge, and Cornell's Animal Science Department offered me domestic rabbits on occasion. Normally, I would have felt remorse sacrificing these animals to use as eagle food, but I had adopted the predatory mind of a parent eagle by then, where food was food with no room for sentimentality. The welfare of my young charges was paramount.

Providing food wasn't the only issue that had become more complicated. Banding the eagles was also trickier since the 1976 technique of putting a yellow Herculite streamer on the right wing of one bird and the left wing of the other couldn't distinguish five birds. We needed a more complex pattern to identify the individuals once they flew. We also had made a mistake using streamers, for we discovered the eagles liked to preen the strips into their wing feathers, obscuring them from view.

With Jim's help and ingenuity, we designed a crescent-shaped tag, also yellow Herculite since that color was most visible on the dark brown plumage. It lay flat on the eagles' feathers and couldn't be preened in between them. By alternating these tags on the right and left wings of P1 and P2 and then gluing red patches on top of the yellow base for M1 and M2, we created a working color combination for all four birds. When M3 arrived later, I chose a green patch for him. The crescent tags could also be distinguished from the 1976 streamers on W1 and

W2 when seen from a distance, identifying the year the eagle had been released. Of course, this all assumed someone would report the tags' shapes and color combinations or at least take a photo of them when an eagle was sighted. We hoped the tags would last for several years so we could keep track of the birds' movements and survival.

The radio transmitters had to be modified as well. Instead of gluing them to the birds' tail feathers (as we had done the summer before), requiring the bird to carry them until the feathers molted a year later, I wanted to find a way to have the transmitter drop off sooner. Ideally, they should lose the radio before migrating out of the area, allowing for long distance flight without the extra weight. I didn't have the technology at that time to follow them far beyond the refuge environs anyway.

After experimenting with several techniques, we ended up sewing the radios into a tiny pouch on their wing tags with cotton thread, which would disintegrate in a couple of months. The radios could drop off by late fall after the birds had left the refuge. At first, we sewed the radio pouch onto the upper side of the tag but soon discovered the weight made the tag slip. By removing the tags and reattaching the pouch to the underside instead, the tags sat snugly in place on the eagles' wings. This second summer of trial-and-error learning was proving to be essential to get things right.

Since none of the birds were ready to fly at the same time, I welcomed the extra help provided by the DEC employees. Although territoriality had overcome me when I first learned I was to be sharing my hillside and my eagles, ultimately it turned out for the best, both for me and my avian wanderers. Peter Nye, Mike Allen, and the other researchers who took turns helping were dedicated observers. I always had confidence they were invested in the eagles' welfare as much as I was. While I chased one hapless eaglet into the cattails, someone else would keep an eye on the others still in the nest. If another took off while I was in the marsh below the hillside, I had to know which direction it flew, or I could be searching for days.

Although they could remain on a chosen perch for an hour or more at a time, the young eagles were rarely idle. This was their time to learn about life beyond the nest. Their heads were swiveling around with every passing bird or aircraft, bending down to watch for fish in the water below, or preening their flight feathers. Occasionally, I could observe two eagles sharing the same perch or log. They would usually ignore one another, acting as if they were alone on their patch of real estate, but every now and then would rub their beaks together, the only indication they recognized one another as nest mates.

When they were back on the tower feeding, territorial behavior increased as food possessiveness became more pronounced. Flapping their wings in each other's faces, snapping their beaks, and blocking access by tenting their wings over whatever they were eating (a common raptor behavior known as "mantling") were all ways they said, "This is my fish, so scram." They were often so much like toddlers fighting over toys that I found myself laughing at their squabbles.

As the eagles developed their flight skills, they would often soar on the thermal updrafts on a windy day. They performed aerial acrobatics with one another, diving at each other in midair, flying above or below the other, or rolling upside down and extending their talons towards their companion. Unfortunately, I never saw them lock talons the way a mated eagle pair can do, but these antics were clearly a rehearsal for later mating behavior when they matured. They didn't like to soar alone and when another eagle wasn't available, I would often spot them riding the thermals with a red-tailed hawk as a stand-in.

The Scout served as my blind once the eagles had decamped to the marsh area. To avoid having them become used to seeing humans, I remained in the vehicle as much as possible when they were perched in trees close to the dike road. Max and Moses assumed their positions in the co-pilot's seat, or in the heat of the day, lay on the side of the road seeking shade from the Scout. I was careful to keep them away from the eagles, knowing that familiarity can lead to loss of fear. They needed to be afraid of dogs as well as humans once they were on their own. I never wanted the birds to take our presence for granted, for once they left the refuge, they would be vulnerable and never again enjoy such protection.

Keeping track of five eagles flying freely was a challenge. My eyes stung from the strain of watching every movement through my binoculars or spotting scope, but I was terrified that I might miss something. It had become more difficult to chronicle their behavior now that I wasn't sitting in front of the video monitor in the comfort of my trailer, a cup of fresh coffee at the ready. I wanted to crawl into the eagles' brains to decipher the intention of every subtle movement, so I could track this learning curve before they felt comfortable enough to leave the refuge.

Caring for birds of prey is the consummate guessing game, especially when they are wild and always at a distance. My interpretation of their motivation and thinking came only from observing their behavior and not from any signals they were intentionally sending my way. Different from my dogs or other pets I have raised, who choose to communicate to people, these birds of prey were self-contained and aloof. It was my job to decipher their code so I could predict their next moves. Admittedly, I wasn't always successful.

---

My penultimate hope, beyond their daily survival, was that I would witness all the eagles catching their own food before they left the refuge for good. In 1976, W1 and W2 had obliged me by exhibiting their fishing skills in the Storage Pool. Although I was devastated that the presence of hunters in the marsh had driven this pair from the refuge, at least I felt confident they could fend for themselves. Once they stopped viewing the tower as their only source of food, I knew they would become self-sufficient.

Bald Eagles are opportunistic predators, resorting to scavenging for food if they need to. Unlike Peregrines and Ospreys, both of whom catch their prey while in flight—the former in the air and the latter in the water—it is not uncommon to see a Bald Eagle wading into the shallows to grab a dead fish or duck for dinner. I had to keep my eyes on the shores as well as on the snags if I wanted to see the birds in action.

In 1977, I didn't have the chance to blame the hunters, for all five eagles dispersed well in advance of the mid-October hunting season. P1, who had never embraced the refuge as home, was the first to leave on August 6, only three and a half weeks after her first flight. This was the same day I released M2 and M3 from their enclosure, although I will never know if P1's departure and the presence of the younger eagles outside the nest were just a coincidence. It seems unlikely that she would be concerned about competition considering her large size and aggressive nature, but maybe she just didn't like crowds.

Her sibling P2 came and went from the tower after she left, but rarely ate there. He departed on August 19, seven weeks after his first flight. These Patuxent birds, being older at the time of their arrival, had always seemed less wedded to their nest site. On the other hand, the midwestern birds—M1, M2, and M3—had been more attached to the nest and the refuge. They fed on the platform for several weeks after release, even when their radios indicated they were venturing into the nearby farmland. They continued to return until late September, when all three dispersed once and for all within a week of one another.

I placed food on the tower until Halloween and checked the nest every few days on the off chance any of the birds might come back. As I climbed the poles with carp in hand, I prayed there would be some evidence of their presence—a piece of food missing or even moved from where I had placed it in the nest—but I was always disappointed. Although I knew their departures were inevitable, I wanted more assurance they were still alive.

Because I never saw the 1977 birds catching food, their tale lacked a solid ending. W1 and W2 had been a success story by comparison. I knew uncertainty was normal and would always be the definition of any wildlife release project. Even after witnessing the 1976 birds' hunting skills, I couldn't be sure they could survive their first winter.

I imagined all the threats that could befall young eagles, yet knew I was unable to do more than hope. I couldn't drive the roads around Cayuga Lake, between Montezuma and Ithaca, without scanning every tree for a large brown silhouette or searching the shoreline for a wading eagle. To this day, I look for eagles wherever I drive, whether on the sides of highways or especially near water, an instinct born of the worry I harbored after my charges had departed.

---

One morning on August 21, with my binoculars pinned on one of the 1977 eagles in the Storage Pool, I noticed another immature eagle nearby with a plain yellow streamer on his left wing. Only W1 and W2 had been outfitted with streamers. At first, it seemed that I was being deceived by my own fantasies. I was so beside myself that I dropped my spotting scope while hauling it out of the back of the truck. With shaking hands, I managed to focus it on the new eagle perched in a dead snag above the cattails. Sure enough, it was W2 who had returned to his marsh a year later. I felt vindicated; he had survived the year and had come back!

All my months of worry and unease, never knowing if my first eagles would survive, were over. By some miracle, at least one had made his way in the world, had caught food, had avoided life-threatening danger, to find his path back to where he had come from. My elation, a mixture of relief, pride, and pure happiness, played itself out that day on the dike road. I only wished there had been someone there to share it with but, as always seemed to be the case during the most exciting moments of this project, I was alone. To their credit, the DEC folks back on the hillside shared my joy when I returned to tell them. For all of us, this was the success story we had been waiting to happen.

For the next month, W2 remained in the Storage Pool, interacting with the 1977 fledglings. I liked to think he was sharing his adventures with them or counseling them on the ways of the world. Occasionally, people spotted him outside the refuge in the rivers nearby and on the northern end of Cayuga Lake. Since I never saw him at the tower or feeding in the marsh, I had to assume he was finding his food elsewhere. He left in mid-September, when the midwestern birds departed, but then returned by himself a week later, remaining in and around the Storage Pool until hunting season resumed the third week of October. As soon as the hunters appeared in the marsh, he took off for good, as he had the previous fall. W1 never joined him that summer, but I would come to learn that their bond, formed early in the Wisconsin nest they had shared, had not yet been broken.

I found it frustrating not knowing where the eagles went after leaving the refuge but comforted myself by thinking it was one of the mysteries of the natural world. Without radiotelemetry, we were dependent on seeing the birds when they decided to show themselves to us. Then again, perhaps we aren't supposed to know everything about the animals we study. Wonder and curiosity keep us from becoming complacent about unexplained behaviors like migration patterns, nest fidelity, and mate selection. We piece together information from sightings and band recoveries, but only enough to whet our appetites for more, not enough to make us stop looking.

---

During the following summer of 1978, after my involvement in the reintroduction program had ended, the DEC staff saw W2 return in July. In August of that year, the 1977 eagle M2, the smallest and weakest of all the eagles released, also revisited the refuge. This report reached me while on vacation in Newfoundland. Even though I was miles from Montezuma, I felt an emotional tug to M2 upon hearing the news. Part of me wanted to be there to see him, to prove to myself he was OK, to witness another success story that my eagles had survived to become independent despite having no eagle parent to raise them.

These 1977 and 1978 sightings of the returned eagles, W2 and M2, proved that Bald Eagles could make it on their own and supported the theory that they often come back to their natal area. This was a great omen for the releases that would follow. When the state took over the program, they could do it with a confidence that belied the many doubts and assumptions that had governed my two field seasons. For me, everything had been unexpected and startling, a revelatory chapter in the annals of eagle behavior.

## — Chapter 21 —

# A Moment in the Sun

When I had first decided to apply to Cornell, I took out the atlas. Upstate New York was the western territories to me, a sweeping unknowable appendage of suburban Manhattan. On the map, the most striking feature was the parallel string of long and narrow lakes that ran in a north and south direction between Syracuse and Rochester. They looked like a gnarled hand reaching out toward the larger Lake Ontario, desperate to spill their waters into this vast reservoir to their north. Their names intrigued me, some difficult to pronounce—Skaneateles, Canandaigua, Keuka—and some bearing names of Iroquois tribes—Cayuga, Seneca, Oneida—that I recognized but knew nothing about. Cornell's famous alma mater was stuck in my brain thanks to my father, who took to singing it to me when I told him I wanted to go to graduate school at this university on the hill:

*Far above Cayuga's waters,*
*With its waves of blue,*
*Stands our noble Alma Mater,*
*Glorious to view.*

The names of the Finger Lakes, so aptly called, would become ingrained in me for different reasons once the eagle project began, but I began to identify the geography of the area as the home of the Iroquois nation.

Even though I was living in their midst at Cornell, I had never had contact with any of their members. I had studied Native American culture as part of my undergraduate anthropology program but was clueless about the Iroquois except as an example of a decentralized Northeastern tribe with a strong influence on colonial America.

My ignorance of this group of people contributed to my dismay one day in July 1977. After a month on Clark's Ridge, I decided to take a short trip to Ithaca to collect my mail and check on my apartment. A couple of DEC folks had arrived the previous night and offered to watch the eagles while I was gone. Leaving them in charge for a few hours felt liberating as I looked forward to my afternoon of freedom. A stack of bills awaited me in my mailbox, however, subduing my mood a bit. I was shuffling through the pile when a handwritten envelope appeared from an unfamiliar return address in Syracuse. Curious, I ripped it open to discover it was from the chief of the Seneca tribe. I knew only that the Seneca were one of several tribes making up the Iroquois nation, but I couldn't imagine why they would be contacting me. Thinking it must be a mistake, I reread the letter several times, trying to make sense of it.

It consisted of one short typewritten paragraph, stating that I had been chosen to be honored by the Iroquois at the New York State Fair in Syracuse that September, one month away. There was a line or two about my graduate research, but this didn't allay my confusion. *Why me? What did this mean? How many others were being honored? Was this a form letter sent to a random group of people?* A phone number was provided in case I had any questions, and they would like to know as soon as possible if I agreed to attend.

I had been raised in a world of thank you notes and RSVPs. Today the computer makes this easy with a click on the REGRET box, but back in 1977, I was forced to pick up the phone. As soon as I heard the receptionist's voice, I began to run through my list of excuses as to why the date of the fair was inconvenient for me. But by the time I reached the third or fourth, she broke in to ask if I understood what an honor this was and to say that if, in fact, I really couldn't attend, they would need to ask the second-place recipient, which was such a shame. She had been down this road before. When I hesitated, she moved in quickly. "I will connect you to the chief, who can fill you in on the ceremony much better than I can."

I was ready to say, "No, don't bother him. I don't really. . . ." but the line was ringing. If only I had had a few more minutes to compose myself, to prepare my regrets—for I was dead sure that this was not something that I could ever do. I had, after all, spent my life avoiding the limelight, and resisted putting myself in a position where I could be scrutinized.

A man's voice picked up—deep and gravelly and "Chief-like," if such a tone existed—and started right off with "Miss Milburn, it is a pleasure to speak with you. On behalf of the Seneca tribe and the rest of the Six Nations, I congratulate you." I stammered and stuttered, trying to convey politely that I had no idea what he was talking about. But then he moved on to my reintroduction of Bald Eagles and the importance of the bird to the Iroquois, "and to all Native Americans for that matter." He spoke of the bird's importance to the traditions and rituals of the Iroquois who consider the eagle a sacred part of their heritage.

My own feathers fluffed a bit as it dawned on me that I was the only one to receive this letter, that I was being recognized because of my work, that there was no way I could say no to this charming man. I would be going to Syracuse. My anxiety soared at that moment even though the fair was some time away and I had no idea what role I was to play.

Many times, during the next month, I was tempted to call him back to say I couldn't do this. But before I reached for the phone, my instincts told me there was something more here. Even though I couldn't see the key to the puzzle, it might be worth taking a chance to find it. I had always been intrigued by Native American culture and its ties to the natural world, and wasn't this a good way to learn more about it? I had only to temper my fears of the unknown.

---

On September 4, shaking with nerves and doubt, I arrived at the New York State Fairgrounds. Hordes of visitors streamed from the parking area, with its acres of white lines and concrete adjacent to the grounds, and merged through the imposing entrance to the fair. While most fanned out in different directions, appearing to know their way around the huge event, a large group assembled in front of the kiosk where an oversized map with a "You Are Here" arrow helped them secure their bearings. I was in the back of the group of newcomers, straining for a glimpse, over people's heads, of the location of the Iroquois Village. Once I navigated my way through the throngs to the overhead archway of the Village, I was directed to the Seneca chief's tent by the volunteer guide standing at the entrance.

When he saw me enter, the chief bowed low, clutched my hand between his two massive palms, his tall velvet headdress festooned with eagle feathers nearly grazing my face. A tunic of soft leather embroidered with multicolored beads fell below his knees. Matching pants and moccasins completed his ensemble. I towered above him from my height of 5′8″, making me feel over-sized but in such awe that it didn't bother me as much as it usually did.

After a few minutes, we were joined by the other five chiefs of the Six Nation Confederacy—the Mohawk, Oneida, Onondaga, Cayuga, and Tuscarora tribes. They all spoke with such reverence about my work that I felt they must think I was someone else.

"Thank you for bringing the eagle back to our people."

"We respect you and your work, which means so much to the Iroquois nation."

"The Bald Eagle is much revered by we of the Haudenosaunee tribe.* We thank you for its return."

Finally, the Iroquois Princess, a stunning girl of sixteen with long dark hair and an elaborate dress of white leather and fringe, who had been chosen to lead the parade during the fair, was introduced to me.† Long tassels hung down from her sleeves and hem, and tiny bells that rang whenever she moved had been sewn into the seams. She wore a simple headdress with one Bald Eagle feather rising vertically in the back. Assigned to be my guide for the day, she was to tell me when and where I was supposed to be.

---

* Native American name for the Six Nations of the Iroquois.

† Every year, a young Haudenosaunee woman was chosen as the New York State Fair's Iroquois Princess. The tradition continues today, but the title has been changed to "Indian Village Princess."

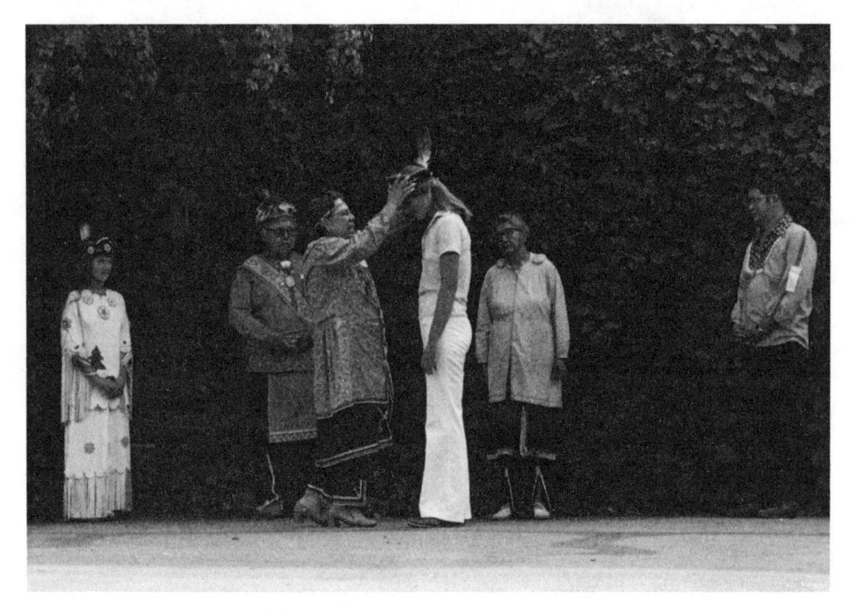

The author receives a ceremonial headdress during the honorary Iroquois induction ceremony. *Photo by Cornell University*

The brief meeting I had expected had turned into a day-long affair, including lunch with all the chiefs and their wives. I have never felt so at sea but also so fascinated by a culture that I knew nothing about. My princess guide was friendly, funny, and was willing to answer all my stupid questions and help calm my nerves. Although I was ten years her senior, her elegance and poise made me feel much younger. With a kind expression and captivating smile, she turned to me periodically to ask, "How are you doing? Are you OK? Is all this overwhelming for you?" Perhaps the emotions that were flip-flopping inside me were visible on my face and worried her.

Once it was time for the ceremony, the chiefs gathered on the stage on either side of me, each wearing his enormous headdress, with me in white chinos and a polo shirt. The Seneca chief started off with a short but moving speech announcing me, the newest inductee into the Iroquois Nation, honored for my protection of their revered Bald Eagle. *Inductee?* He went on to explain I was now an honorary Iroquois and had been given the Haudenosaunee name of "Yea-Non-Dea" meaning "She

The author celebrates with the Iroquois wives and Princess after being inducted into the tribe. *Photo by Cornell University*

Who Feeds." Seemed quite apt, and I didn't even need to remember it because he handed me an official ID card with my Iroquois name on it.

Unexpectedly, the wives of the chiefs then came forward while the men retreated to the back of the stage. They formed a circle around me and showered me with gifts, all representative of their handmade crafts. Each tribe had its own specialty, and my arms were soon laden with beaded jewelry, straw baskets, corn husk dolls with hand-sewn felt clothes, beaded bags, and porcupine-quill necklaces. One of the ladies placed a dark blue velvet headdress on my head with a lone Bald Eagle feather in the back. Since I was so much taller than she was, I had to bend down for her to reach me. As this was going on, the young Princess was dancing to music and waving her arms with the bells ringing all the while.

After the gift giving, I thanked and hugged each woman as the men shyly stood by, a respectful distance away. It baffled me why the chiefs had taken a backseat at the ceremony when earlier it had been obvious they ran the show. As I prepared to leave, the wife of the Seneca chief pulled me aside. She whispered into my ear, apparently not wanting anyone to overhear.

"Thank you for coming today. This was our very first ceremony, all made possible because of you. Traditionally, our honorary Iroquois has always been a man, so women haven't been able to participate in the event. This was a special day for us. You saved the Bald Eagle and allowed us, the wives of the chiefs, to celebrate you. We thank you." Then she placed both my hands between hers. Because I was female, this ceremony had been as much of an honor for them as for me. Tears welled in my eyes as I realized the significance of this moment.

On my drive home, I mulled over the events of the day—what I had learned about these people and their culture, and how much I still had to learn. I was proud of the work I had done and how much it meant to them, but also would never forget the pride on the faces of the wives who had yearned for and finally received what I had long been avoiding—their moment in the sun. The women's radiance and happiness told me what they must have been feeling. While I might have been conscious of my gender as I navigated the world of raptor research, they had broken their own barrier today in a ceremony previously restricted to Iroquois chiefs. I knew that this day in September 1977 would be a new beginning for them as well as for myself.

## — Chapter 22 —

# The Waiting Game

By late August, once the five eagles had left their nests and were flying on their own, the DEC employees returned to their offices. The hillside felt quiet and deserted like a once bustling stage set after the curtain drops for the last time. P1 and P2, having left the refuge weeks earlier, never returned.

I remained at my campsite, keeping vigil into late September, when W2 and the three Midwestern birds—M1, M2, and M3—stopped coming for food, instead remaining in and around the pools. Although I continued to put fish on the tower, the birds didn't spend time there anymore. Instead, they would swoop down, grab the fish with their talons, and carry it toward a sturdy snag in the marsh to tear apart their meal in private. Some days, the fish remained untouched, leading me to believe they were catching their own food, a reassuring sign. As in 1976, I found that interspersing the fish with pheasants and rabbits enticed the eagles to feed on the tower more often. They probably knew they could catch carp on their own but enjoyed the variety that would sometimes magically appear in their nests. I often felt like a parent whose teenaged children were just beyond reach, stopping home only to drop off their laundry and grab a snack. The snacks had better be good, or they might stop coming.

I finally had to admit to myself that my round-the-clock presence on the hillside was no longer needed. Classes at Cornell had started after Labor Day, so I had already missed the first week. I drove down to Ithaca one morning to explain to my professors that I might be delayed: "I hope it's OK if I join class late, but my eagles might still be in the area." They were all understanding but let me know if I missed more than one week, I would get so far behind I might have to drop their course. Reluctantly,

I realized it was time to break camp and return to my Ithaca apartment. I could continue to oversee the project from there by traveling daily to the refuge. Packing up all my belongings, arranging with DEC to have the little trailer hauled away, and cleaning up the site that had been home for two summers—my preparations for departure were bittersweet. Unlike an eagle parent who can decide either to follow their young or allow them their newfound independence, I had no such choice.

Unlike the first year when I departed, filled with delight at the idea of returning to civilization and the land of hot showers, this year's leaving only saddened me. The fact that I would never again be coming back to live on Clark's Ridge diluted any relief I might have felt at going home. After two seasons on the hillside, I had become used to spartan living and my outdoor accommodations. The campsite had become home and even though I would be happy to see my bed and apartment again, reunite with my friends, especially Larry, who had returned from his summer research by then, I had a hard time saying goodbye. Even Max and Moses looked downcast as I loaded the Scout with all traces of our presence. Perhaps they were worried they were going to be left behind, like most dogs when they see suitcases appear. They couldn't know they were going to be part of my backseat research team since I would be commuting from Ithaca to the refuge for another month.

I drove the nearly sixty minutes up Cayuga Lake to Montezuma each day. The eagles had gone, but if they came back, I had to be there. After a few days of radio silence, as I wandered along the dike roads and traversed the farmland, scouring every inch with my binoculars, my vigilance was finally rewarded by the sight of W2, back in the marsh, preening himself on a dead snag. What a welcome sight he was, confirming my decision to check in daily. He had proved his site allegiance was strong and indicative of his success finding food in the area. He had come back alone this time without the midwestern birds and remained in the Storage Pool until hunting season began in mid-October.

With classes having resumed by then, I was juggling course attendance with my commute to the refuge. Some days, I was at Montezuma before dawn, in position on the dike road alongside W2's favorite section of the pool in case he made an appearance. I would watch him all morning and rush back to campus in time for my afternoon classes. Occasionally, my class schedule was too tight, and I couldn't make the trip until the end of the day, forcing me to drive back at night. I hated watching the days shorten as we moved into October, when darkness often descended while I was still on the marsh. Fatigue caught up to me on the lonely drives home along the lake. On more than a few of my drives down Rte. 89 to Ithaca, I worried about falling asleep at the wheel or, equally frightening, hitting a deer, many of whom bounded across the road in front of my headlights.

It was a frenetic time of trying to be in two places at once, but I knew it wouldn't continue for much longer. I was facing the end of my time in the refuge, which would come as soon as W2's signal was gone. He was the last of my eagles to remain, ironic considering that, with his sibling W1, he had been the first to arrive at Montezuma the year before.

When driving north to the refuge in mid-October, I followed the lake's shoreline, scanning it as always for an errant eagle scouting for fish. No signal had come from W2 in the week after the hunters' guns began to sound in the pools. It seemed evident that, just as in 1976, he had been driven away by the noise or presence of people sharing his hunting area. I continued to check on him for the next week on the off chance he would return. To ensure the food left for him was always fresh, I cleaned out the nest and provided new fish and other items every day. But the food remained untouched, there was no signal from his radio, and I didn't spot him anywhere. It was time to admit that he had left for good.

---

My final departure from Montezuma at the end of October 1977, down the dirt road leading away from Clark's Ridge, brought back memories of the first crates of Bald Eaglets arriving at their new home. I had had no idea how my life would change as the birds were unloaded from the trucks.

Departures often cloud the challenges we leave behind until all we can think about are the good times and successes. While my job raising seven young Bald Eagles was done—I could do nothing more to protect them or ensure their survival—I was saddened to think I wouldn't see them again. During the weeks following my return to Ithaca and my life as a student, I fell into a melancholic state. For two years, my purpose had been clear but was now muted by the eagles' final exodus. I had to redirect my energy to classes and thesis-writing, neither of which could supplant the adrenal rush I had felt on that hillside. Caring for animals had been an important part of my life since childhood; I wasn't going to fill that niche easily.

I hoped sighting reports of immature eagles with yellow tags would keep me informed of their movements, but I also knew that it would be New York's DEC employees who would now be privy to this information. From the start, my future assignment had been made clear: write my thesis, outline the hacking techniques we used so others could follow them, and step aside to allow the state to take over the program and release more Bald Eagles into the sky.

---

For the next forty-five years, I never returned to Montezuma. In the ensuing decades since leaving those marshes, I traveled the New York Thruway several times, on trips out west with my family and visits to the Great Lakes. My eyes always sought out Clark's Ridge, my remote home for two summers, to see if there was still a tower up there or even better, to see an eagle nest high in a tree near the marsh. A permanent break from a project that had been so emotionally and physically intense seemed the right move. I couldn't be a mere observer after being so much a part of the action.

Knowing how much care and attention it had taken to raise just seven birds, I admit to being concerned when I heard about the state's intention to release eagles in far greater numbers in the years after 1977. The success of those first years seemed to have given DEC the confidence that more would be better in restoring the Bald Eagle population and certainly, from a scientific standpoint, this was true. To increase the chances that the eagles would spread out and reproduce throughout the Northeast once again, the skies had to be flooded with young birds. But I had a hard time letting go of my focus on the individual in favor of the larger population. The time had come for me to leave this project to others.

# Part Four

# Beyond Eagles

## — Chapter 23 —

# The Next Chapter

After my last eagles left Montezuma in October 1977, I felt unmoored. Although I had immediate tasks waiting for me, a void loomed beyond them. I had followed—admittedly, not always in a straight line—a paved path to my chosen destination for so long that I didn't know what to do once it ended. For me, the release project was over, I had done what I came to Cornell to do, and now I had to write the next chapter of my life. To make things more complicated, my decisions no longer affected just me.

Larry had left Cornell by then, his doctorate in hand, to start his first job in Ipswich, Massachusetts, at the Quebec-Labrador Foundation, a nonprofit environmental conservation group. I spent that winter completing my master's thesis in his apartment in Portsmouth, New Hampshire, a fifty-minute drive from Ipswich, on the shores of the Piscataqua River dividing New Hampshire and Maine. While I hammered out pages on my typewriter, tugboats, barges, and container ships navigated up and down the river, pursued by cormorants, gulls, and diving ducks, searching for dinner in the boats' wake. Today, Bald Eagles would likely have been right behind them, but in 1978 their absence confirmed the significance of the work we had done.

Writing my thesis, as difficult as such tasks always are, allowed me to pull together the disparate parts of my research. Those long hours spent in the blind observing and categorizing eaglet behavior paid off when I compared my data to the documented studies done on wild young. My birds provided evidence that being raised in the absence of parents was not detrimental to their normal way of doing things, including finding food and breeding. This was an auspicious sign for raptor conservationists, who were now able to view the hacking program as the eagles' salvation.

My thesis also served as a recipe for hacking, valuable to DEC in 1978 and to the many other states who adopted the techniques after that. I was fortunate to have the Peregrine releases as a model and needed only to tweak the procedure to fit the Bald Eagle. Although I saw our work in 1976 and 1977 as miraculous on many levels, it may have been preordained when US Fish and Wildlife decided to offer the program to Tom Cade. He and his team held the keys to success.

As I wrestled with mounds of data, statistics, and "notes from the tower" that fall, Larry and I decided to be married after my thesis was complete. In May, I defended it in front of my committee of professors—a nerve-racking experience but not nearly as bad as the anticipation. I knew my subject, and my teachers wanted me to succeed. After packing up my apartment in Ithaca, I drove home to Long Island with the finished manuscript on the passenger seat, my dogs and belongings in the back. Our wedding took place two weeks later. We spent our honeymoon in Newfoundland where we immersed ourselves in the wonder of seabird colonies and whales, Newfie music and culture, and the barren yet bountiful landscapes of the province. It wasn't until we returned to New England a week later that the whirlwind, which had defined my life for the past few months, began to slow down enough for me to think about what lay ahead.

After so many years of struggling to get to this point, I faced a frightening uncertainty and wasn't mentally prepared to leave the haven of the academic world. When I had arrived at Cornell three years earlier, my goal had been to study for a PhD. I was halfway there, but the rest of the journey seemed to stretch endlessly before me. I wasn't ready to say goodbye to eagles—they had become a vital part of me over these past two years, my connection to them a life passion—but I knew to study them I would need to leave the Northeast and find a locale with a viable population. Larry had just started his job and needed to be in Massachusetts for the foreseeable future.

A biology professor at Virginia Tech in Blacksburg invited me to study Bald Eagles in the Chesapeake Bay area as part of a three-year research project he was conducting. This would have been a behavioral study, purely academic and meant for the advancement of science and a doctoral degree, not the hands-on work I had been doing. But it was an eagle study, nonetheless. While part of me fantasized about climbing up to eagle nests to band young, following adults throughout the Bay's inlets and up rivers, and perhaps becoming one of those eagle experts I so admired, another part—the more idealistic part—wanted to effect change by making life better for wildlife. I had entered the world of bird conservation to help species survive, not merely to observe their wild behavior. But a more practical reason also guided me.

We had just bought a house in Newburyport, close to Larry's office, where I was looking forward to settling into our first home. Some tough decisions faced us during this time, similar to those faced by many couples with dual career interests. Larry's job was a tremendous opportunity for him, while my future was cloudy at best. It seemed natural that I be the one to be flexible given that we wanted our lives to be together under the same roof. Hubris engulfed my thinking, and I believed if I could succeed with eagles, I could employ my bag of tricks anywhere and be greeted with open arms. I hoped to have it all.

I only needed another species, closer to home, one that was declining and could benefit from hacking. A nonraptorial bird would allow me to prove the technique could be adapted to species other than birds of prey. I was back on Tom Cade's doorstep, asking for ideas. He suggested the Loggerhead Shrike, a bird he had studied earlier in his career, before entering the world of Peregrines.

I read as much as I could on the shrike, since I knew little about the bird. Studies revealed a fascinating species that is a predatory songbird, not a raptor. It has no talons, but has a tiny, hooked beak, and it is smaller than a robin. It kills its prey—usually lizards, small rodents, or large insects like dragonflies—with its beak like a raptor but then impales the carcasses on thorns or barbed wire to eat later. Most importantly, the shrike was in trouble, listed as threatened in most states and had disappeared from its former breeding sites in New England.

Enthusiastic about modifying hacking to accommodate the bird's unique behavior, I launched a research proposal, convinced that support would pour in from every corner. I could live in Massachusetts, hack the shrikes back into their former habitat, and earn my PhD from Cornell. To me, it all made sense, and Cade was on board with the plan, assuming I could secure funding.

The returns proved otherwise. I was rejected by each funding agency I approached, always with the same devastating response. All agreed the Loggerhead Shrike was a fascinating bird and was in decline—mostly because its habitat of open fields and farmland was disappearing. But the species, whose nickname was the "Butcher Bird," didn't have public appeal. Most people couldn't even identify it, much less care about it. Each funder tried to placate me by saying, "We would be happy to fund you if you were to stick to Bald Eagles."

It was time to face the harsh truth: I was not the mainspring of the eagle's success but instead, by doing the heavy lifting, had attached myself to its splendor and celebrity. The eagle's own charisma, together with the nation's patriotic need for its national symbol to endure, had made it all possible. My success was tied to this bird and without him, my options had limits. Contrary to the belief of the newspapers back in 1976, the real story was the Bald Eagle, not me.

A self-defeating part of my nature apparently requires hitting a wall before I can muster the energy and chutzpah to change direction. My undergraduate career was certainly proof of that as I shifted from one major to another, from one campus to another, from one career ambition to another, all motivated by an event, an epiphany, or an unwelcome happenstance. My rejection by funding agencies precluded my shrike research ambitions but stimulated me to regroup, change course, and refuse to wallow in self-pity. I wouldn't be honest if I said I didn't wallow a bit, lamenting my fate at the hands of those who placed politics before conservation, but I knew that it would get me nowhere. As with all my other experiences, luck and timing played their role in my next adventure.

---

Although my plans to do field research and hack young Loggerheads had unraveled, a piece of the project survived. The US Fish and Wildlife Service contracted me to do a breeding survey of the species in the Northeast Region, extending from Virginia to Maine, to confirm whether the Loggerhead had declined enough from its original population to warrant listing as a threatened or endangered species. To do this, I needed to track down all the historic records of nest sites in each state and determine which were still active. Traveling to regional libraries and pouring through museum documents and state bird guides was dull compared to the excitement of working with live eagles, but my wonky academic self didn't mind the research, even if I recognized it would be short-term and not lead me to my coveted PhD.

As I plowed on through the shrike nesting data, a friend sent me a job posting with The Nature Conservancy (TNC) that sounded tailor-made for me. TNC was looking for someone who had knowledge of wildlife biology and endangered species to initiate their Natural Heritage Inventory Program in New Hampshire, just north of Newburyport.

Biologists at TNC created the Natural Heritage Programs in 1974 to identify and catalog the biodiversity of the fifty states. The intention was to map all elements of our natural heritage—animals, plants, and habitats—thereby protecting them by knowing where they could be found.

At first, I couldn't grasp the value of such an inventory, which seemed labor intensive and mammoth in scope. But then I thought about it all in terms of the Bald Eagle. If a nest was documented, its location would be accessible to both land conservationists and developers. Any building project—including housing subdivisions, industrial complexes, highways, and shopping malls—would have to be viewed with consideration given to the nest site and critical area surrounding it. This seemed like a reasoned approach to wildlife habitat protection.

In the late 1970s, when I applied for the job as New Hampshire Natural Heritage Director, programs had been launched successfully in almost every state, and TNC's eyes focused on New Hampshire as one of the last holdouts.

For my interview, I flew down to the main office in Rosslyn, Virginia, just outside of DC, and met with all the directors of the various TNC departments. It was a Friday in 1980, a day I remembered only because after many interviews, I was invited to join the staff at the office's TGIF party in one of the conference rooms. There were few if any women invited to the table during this time, which was in keeping with the environmental world in general. I felt fortunate my game was in New Hampshire, where the playing field was lower key and more forgiving than in DC.

---

Although my eagle project seemed to intrigue my TNC interviewers, surprisingly it was my Loggerhead Shrike survey work that won me the job offer. I understood the importance of inventories, of keeping records of breeding sites, territories, and migratory routes of wildlife. The Heritage program was essentially a database that did just that.

Stunned to be hired so readily, I said yes without blinking, unaware what the job would entail. Because I hadn't finished my shrike survey yet for Fish and Wildlife, I was going to have to juggle both contracts simultaneously. Overnight, I had gone from field biology into something more akin to desk biology, whether it be researching shrike documents or crisscrossing New Hampshire to track down environmentalists to bring on board with the TNC agenda. In between writing progress reports, I was learning about the sector of conservation work that nobody talks about in graduate school, where our eyes are blinded by visions of rivers and mountains, animals, and open spaces.

At times, I was frustrated not to be using my training in animal behavior and missed the natural world of eagles and the refuge. But I also knew that I had chosen this more predictable lifestyle and had to forge a different path. Working for Fish and Wildlife and TNC offered me a preview of what opportunities were out there for me.

Both jobs contributed to my natural history knowledge. When driving back and forth to New Hampshire for meetings, I took detours to natural areas and sanctuaries to look at birds and discover what conservation projects were being conducted. Although I enjoyed learning about the Osprey and loon work on the lakes, both efforts reminding me of my eagle work, it is telling that I spent a year traveling throughout the state without seeing a Bald Eagle. This would be unheard of after the year 2000, when the birds had begun their dramatic comeback, but back then my eagles hadn't had time to restore their presence in New England.

I had been told by TNC to "sell" the Heritage program to the state, an unlikely task. Instead, I donned my research cap and approached the local environmentalists with binoculars in tow, eager to learn about the natural history of their state. I may not have been as forceful as TNC wanted, but it proved the best approach, at least at this stage of the program.

The most successful encounters I had involved walks and talks. I found that New Hampshire people have a strong affinity with the outdoors and love to share it. Trekking around a lake or climbing a peak in the White Mountains would lead to good conversation and honest opinions. I could drive away knowing I had connected; if not on a professional level, at least on a more important personal one.

By the time my TNC contract was up at the end of the year, I was ready to move on. The last few years had been a lay-by in my life as a wildlife biologist, and I was antsy to be outside and working with animals. After struggling to be accepted into the club, it had become frustrating to be on the outskirts, looking in at the work of others. With the shrike survey and the Heritage program, both worthwhile in their individual missions, I had become separated from hands-on field research and wildlife protection. The time had come to get back in the action.

# — Chapter 24 —

# A Second Refuge

Determined to return to the world of wildlife once my contracts with TNC and Fish and Wildlife neared their end, I anticipated it might take some time to find my next job. Fortunately, I had stayed on the mailing list of several zoo publications after my stint in Louisville, and the announcement of a job opening for Curator of Birds at the Franklin Park Zoo, Boston's largest zoological park, caught my attention. Yes, I had left zoo life behind to go to Cornell, and no, I hadn't forgotten my reservations about captive animals and their treatment. But I thought being a curator might allow me more control, and perhaps I could make a difference.

The two-hour drive, in heavy traffic, from my home north of Boston to the zoo on the south side of the city should have been a red flag. At that time, Franklin Park was ready for an overhaul (it was eventually renovated in 1984), its age and outdated enclosures disheartening. I had tried to push the negativity I felt toward zoos aside, but as I walked past the mammal enclosures, most of which were even smaller and more confining than those of Louisville, I realized there was no going back. Although the birds in the spacious aviary appeared to be well cared for, I wouldn't be able to disregard their less fortunate mammalian neighbors or ignore my visions of free-flying Bald Eagles and Peregrine Falcons. After my interview, I drove back through Boston, again in rush hour, and knew I wouldn't take the offer if it came. It was fortuitous that I didn't; three weeks after my zoo visit, I found out I was pregnant.

Larry and I both wanted a family and were aware that the timing on this might not be up to us. I hadn't expected things to change so quickly, for the brakes to be put on to my career journey so jarringly, but I honestly couldn't wait to see what this new adventure would bring. I never suspected the next decade and a half would require so much creativity and resourcefulness to balance family life and career, ensuring one was not forever infringing on the other.

My shrike status survey remained unfinished, requiring me to spend the winter and spring tying up the loose ends. Our daughter Whitney was born in June 1981, before the final report was complete. I accomplished this task during her naptimes when my eyelids wanted nothing more than to close so I could join her in blissful slumber. Eventually, my paid employment came to an end, and I transitioned into my new maternal role, perhaps not as financially lucrative, but rewarding even so.

---

From the birth of our first child in 1981 to our fourth in 1989, I vacillated between a preoccupied and zombie state, depending on my degree of sleep deprivation. Four children soaked up my wellspring of maternal affection as much as seven eagles had done a decade earlier. At that time, women were still reeling from the influence of Betty Friedan and Gloria Steinem, trying to balance career and family, logistically as well as emotionally. Guilt was pervasive; no matter which way we turned, no glass slipper fit all of us perfectly. I had studied to be a field biologist, requiring long hours observing wildlife, usually during the warm months, and usually away from home. With four children whose needs ranged from infant care to homework help to attendance at soccer games, this career path was no longer sustainable. As with the eagles, I resisted delegation. I enjoyed being home with my kids but also missed the work I had been trained to do.

Meanwhile, sundry activities filled the void. Joining the board of a regional land trust and the town conservation commission, and teaching nature study in the children's classrooms kept me busy and free to respond to emergencies on the home front. Working as a real estate broker and consultant for environmental nonprofits—a juxtaposition if there ever was one—both taught me about land use planning and helped buy new shoes, the only items not lending themselves to hand-me-downs. It wasn't perfect, but by cobbling together a home and work life, I felt fulfilled, until I wasn't. Even as the children became more independent and immersed in school activities, however, they still needed a parental touchstone. I knew the time had to be right before I could return to a career in biology.

---

After living in a small house in the center of Newburyport for four years, Larry and I yearned for open space. It soon became evident that neither of us was an urban dweller.

In Rowley, a rural town a few miles to the south, we bought a more spacious house—spacious because it was a renovated barn—boasting eleven acres of swampland and pine forest. Before the coffee maker and plates were unpacked, we had already installed wood duck boxes, bird feeders, a boardwalk through the swamp, and a swing set. With us came our two dogs, Max and Minke (Moses had passed away by then), our neighbor's cat Blossom, who had adopted us before we moved; our eldest daughter Whitney; and our second daughter Alix, not yet born but seven months along in my belly.

For the next fourteen years, we enjoyed our house, despite the constant home repairs—it really should have remained a barn—and healthy population of mosquitoes. During that time, we welcomed two more children to our family—daughter Casey and son Michael. The house's open spaces and gardens served our expanding group well, until the bulldozers arrived. Rowley had been slow to institute zoning laws. Developers had set their sights on those rural towns of Massachusetts' North Shore where real estate was still cheap and town boards naive, tax-greedy, and oblivious to the effect unbridled building could have on a town's character and natural areas. It was time for us to move again.

Whenever we ventured south to visit family in Connecticut or New York, we took a shortcut to the highway through the town of Boxford, where we admired a property marked by a hand-painted sign hanging over the driveway: "Witch Hollow Farm." The name made it a favorite landmark for our kids, while Larry and I were taken in by the white Georgian farmhouse topped by massive black chimneys, its gigantic red barn, and most of all, the expanse of open pastureland surrounding the buildings. My love for farms hadn't diminished since my childhood, and I always hoped someday to live in such a place.

In 1995, Witch Hollow went on the market. The Trust for Public Land, a national land conservation group, had been enlisted by the town to purchase the farm from the former owners, who were planning to sell it for development. The property had sat vacant for over a year, potential buyers scared off by its general state of neglect. Paying no heed to such mundane concerns, we bought it, turning ourselves into farmers, fixer-uppers, and wildlife custodians overnight.

With its dual ecosystems of farm and forest, Witch Hollow became a successful blend of man-made and natural habitats. The only separation between the domestic animals and wildlife was in the former's dependence on us for their safety. Fences, daily rations, fresh water, medical care—these were our gifts to the farm life. We couldn't protect the wild creatures from the natural toll of disease and predation, but by managing their habitat, we could at least ensure they had a home.

---

The house, built in the early 1700s, was in dire need of TLC, and the 1850s barn, once home to livestock, wagons, and hay, now stood empty and forlorn, awaiting its next chapter. When Larry first saw it, his comment, "It smells too clean. It needs manure" proved prophetic.

The following summer we took the kids on a sailing trip to Cape Breton, Nova Scotia. One late afternoon, we passed a hillside pasture where several head of Scottish Highland cattle were grazing, a picture postcard moment. Before dawn the next day, leaving the rest of us still asleep in our bunks, Larry rowed over to the shore of the farm, found the farmer in his kitchen having breakfast, and offered to buy two cows. When he returned to the boat, he announced, "I found the solution for the barn!" Luckily, even he knew how crazy it would be to transport cows from Nova Scotia to Massachusetts. Instead, the farmer had given him the name of another Scottish Highland owner in Rhode Island. Wasting no time and afraid I would sabotage his plan if he came back to the boat before closing the deal, he phoned the American farmer from the Canadian farmer's kitchen.

Scottish Highland cows are the icons of choice for every billboard and poster advertising Scotland. Their shaggy reddish-brown hair and long curved horns connote Highland flings and fields of heather and gorse stretching to the horizon. Raised for beef, they are hardy for any type of weather we have in the United States. They could stay outdoors all year, and supposedly wouldn't need any care other than providing them with hay and water. More docile than our American black angus, they should be relatively easy to handle. At least that's what we were told.

When the Rhode Island farmer delivered our two "calves" in a trailer, they told a slightly different story. They were half-sisters, six months old, one the traditional reddish-brown with perfect outwardly curving horns, while the other had the more recessive white coat with two misshapen horns—one heading outward and the other twisted downward until it curled back towards her cheek.

Our neighbors had assembled for the arrival of the calves. One woman, an animal owner herself, walked right up to the calf with the cockeyed horn. The "docile" new member of the family lowered her head and charged, hitting our friend square in the stomach and lifting her off the ground, instantly dispelling the myth created by the farmer that these were "child-friendly pets, as tame as dogs." That night we decided on their names—Thelma and Louise—appropriate monikers named for the notorious outlaw duo of Hollywood fame.

Thelma and Louise lived long lives—much longer than most cows—and died peacefully at age twenty-two and twenty-six respectively. Our town mourned their passing. Everyone who drove past the farm looked for the pair, referring to them as "bulls," "yaks," "buffalo," but never Highland cows. They were an anomaly for Boxford and provided passersby with a peaceful pastoral scene they could enjoy during their daily commutes or on weekend excursions with their children to see Thelma and Louise grazing by the side of the road.

Although not as majestic looking as the cows, two mini donkeys joined our small livestock collection. We inherited Anthony from our mortgage officer who was moving and couldn't take him with her. With Anthony came a mini horse, Spudzi, his best friend; the two were a package deal. Would we mind giving them a "temporary" home? Luckily for them, I had had donkeys when I was young and knew what I was getting into.

When I was ten, my godfather, who owned a horse farm in South Carolina, gave me two baby mini donkeys for my birthday. They were grey fuzz, as tall as my waist, with muzzles of pale velvet, tiny feet that could have fit into Friskies cans, and marked with the black cross, assuring me their ancestors had carried Mary to Bethlehem. I never held much for the word *temporary*. Donkeys have a way of nudging their way into your soul. Unlike other pets whose lives are too short and rip your heart apart when they leave, donkeys are with you for the long haul.

The cleverest of all large farm animals, donkeys match the intelligence of dogs more than cows, sheep, or horses. When Anthony arrived, we were warned that he was like Houdini in his ability to escape, and he proved his talent many times. All the latches on his stall doors had to be tied with rope, or he could jiggle them open. It was a hard and fast rule of the farm to never go in or out without locking the door behind you. Luckily, they both loved food, especially Fig Newtons, so they were easy to lure back into their stall if they got past us, assuming we knew they were out.

Driving home from work one day, I saw a line of cars backed up along our road. The cause of the traffic jam was in front of our driveway: Anthony and Spudzi happily eating the tall grass along the shoulder with their rear ends projecting out into both lanes, oblivious of the stalled cars with their amused but impatient drivers. A policeman was trying to get them to move by pulling on a rope he had put around Anthony's neck. Mules and donkeys deserve their stubborn reputation. Anthony was ignoring the officer completely, and when the rope was pulled hard, he sat down in the road on his haunches. The poor man, turning red with frustration, resembled a character in a *New Yorker* cartoon as he tried to pull Anthony up from a sitting position.

Feeling like Superman, I grabbed the grain bucket and a couple of Fig Newtons, and had both animals trotting along behind me in seconds. Apparently, Anthony had worked the latch on the outside gate to open it right onto the busiest road in town during the busiest time of day. The event not only made the police blotter but also the front page of the paper.

Spudzi had been with us for eight years when he contracted an aggressive brain tumor. A friend had warned me that donkeys, herd animals like cows, hate being alone and become anxious when they lose a partner. Sure enough, as soon as the vet arrived to put Spudzi to sleep and remove him from the pasture, Anthony went berserk. Galloping at top speed, trying to scale the stone walls separating him from the road, hee-hawing in a frenzy that made him seem unhinged, he was letting us all know he was not going to be left alone. I put in an emergency call to a woman in the next town who owned a herd of donkeys and asked if I could borrow one on short notice.

A neighbor with a horse trailer came to my rescue, driving me to pick up a little female, a ten-year-old named Legs. I expected her to be the answer to Anthony's dreams, but like many budding relationships, there was a tense getting-to-know-you period. After a dramatic week of kicking and biting his new companion, Anthony asserted himself as boss of the pasture, a truce ensued, and the romance of the century began in earnest.

In addition to donkeys and cows, we found ourselves awash in animals of all varieties, both domestic and wild. We had intended to provide a haven for wildlife but soon chickens, quail, dogs, and cats as well as a plethora of deer, coyotes, foxes, fishers, raccoons, possum, various rodents (both the good and the bad of the kingdom), an occasional bobcat, snakes, turtles, and many species of birds frequented the fields and woods of our land. All the domestic animals, except the dogs and cows, were orphaned, the group growing as people heard about our propensity to adopt anything needing a home.

We had to keep the chickens penned in an outside enclosure that resembled Fort Knox. One summer, a vixen decided to make her den under the chicken coop, and as the hens pecked for grain on one side of the wire, the baby foxes cavorted on the other. Besides falling prey to the occasional long-tailed weasel, adept at either squeezing through gaps in the boards or using rodent burrows to emerge into the coop from below, the chickens remained unharmed for the most part, a miracle considering their presence in the hub of predator nirvana. Coyotes, foxes, and fishers steered clear of the pastures for the most part, probably due to the presence of Thelma and Louise with their menacing horns. The Canada Geese who nested on the edge of the pond in the cow pasture depended on their bovine neighbors to protect their goslings. The donkeys served as sentries, hee-hawing at top volume if they saw anything nearby that looked suspicious. If I was late putting them into their stalls at night, they let me and the rest of the neighborhood know, so great was their fear of having coyotes invade their field after dark.

---

Although we owned only a few acres, we leased the remaining conservation land from the town to manage for wildlife. Our first years on the farm were spent creating a sanctuary for those animals who had once made their homes here but had been displaced over time. We planted specialized perennial grasses in the fields to encourage grassland birds to nest, including Bobolinks, Eastern Meadowlarks, and Savannah Sparrows—species at risk due to habitat loss—and seeded meadows with wildflowers to attract pollinating insects. By supplying bird houses for cavity nesters like Eastern Bluebirds, Tree Swallows, House Wrens, Screech and Barred owls, and Kestrels, we offered substitutes for the dead snags they required for nest holes.

Working with the Fish and Wildlife Service, we erected bat roosting boxes in the barn so bats could winter over and avoid encountering white-nose syndrome (WNS), a fungal disease that decimates bat populations, especially those seeking refuge in caves during the cold months.

Often the best management technique was not to manage. For deer, we left sections of fields uncut to provide sleeping and hiding areas during the winter months when coyotes were a threat. For Chimney Swifts, declining due to lack of nest sites, we safeguarded access to our chimneys for their nests, by not capping the tops and never building fires during breeding season. In the garden, I encouraged wild plants like milkweed, so essential for Monarch butterflies, and even tolerated the chipmunks, woodchucks, and deer who "shared" my vegetables. That doesn't mean I didn't yell at them the next morning when I saw bites taken out of my tomatoes and squash, however.

Hawks and owls thrived in the open habitat. Thanks to a barn filled with grain for donkeys, cows, and chickens, a bounty of small rodents provided a feast for the red-tails and other avian predators. At dusk in early spring, the nasal "peent" calls of displaying American Woodcocks at the edges of the pastures and the Barred, Great Horned, and Screech owls calling for their mates, joined the evening symphony of frog choruses, emanating from the ponds and streams dotting the farm's landscape.

Our weekends were spent managing the land for wildlife but also for the public. In return for being able to lease the property for ourselves, we created and maintained an extensive system of public nature trails used for walking, exercising dogs, and birdwatching. We were surrounded by acres of woodland, stream crossings, wetlands, and meadows. How to design trails to provide access to all this without cutting down trees or destroying wildlife habitat was a trick. Our neighbor Ralph Abruzzi, an eccentric but generous soul who had roamed the woods as a child, became indispensable to our trail efforts with his knowledge of plants, habitat, and wildlife needs. Ralph would walk into the forest with a chainsaw and clippers and return several hours later, having designed a path that left every tree over three inches in diameter intact. His feel for the land and its contours was intuitive.

My wish to reconnect with animals and to use my education to help species survive was granted by life on the farm. I may not have had Bald Eagles to enjoy, but we had created an oasis that catered to wild creatures, where we could meet the needs of those whose habitats were in jeopardy. Our home was a wildlife refuge, and we were its managers.

## — Chapter 25 —

# From Field to Classroom

*First day of school. September 1997. Seventeen pairs of eyes boring into me. Holding myself erect, I tried to project the in-charge demeanor I knew teachers should convey. Everyone was wary of this stranger in their midst. I wanted to say, "Hey, I'm just as worried as you are, about how we're going to get through this course in one piece, about all the long unpronounceable words I have to teach you." Instead, my stomach churned, my kneecaps trembled beneath my frumpy teacher slacks, my voice wavered as I babbled on, terrified of silence slicing through the classroom air. I don't know when their eyes began to glaze over. Maybe when I handed out the leaden textbooks with the dragonfly on the cover and asked them to turn to Chapter One. A rookie mistake.*

My segue from research to teaching was as circuitous as my other life directions had been. Although our farm kept me directly involved in wildlife protection, when my youngest child approached school age, I wanted something more challenging. Knowing my background, one of my daughter's teachers recommended I try tutoring at a private high school nearby. Baby steps always worked best for me. If someone had asked me at the time if I wanted to teach, I would have responded with an emphatic no. The idea of standing before a class, spouting pearls of wisdom, didn't appeal to me. But one-on-one tutoring might be a good beginning.

Basic science proficiency, so confounding in my early days of graduate school, had eventually nested in my brain. While empathizing with students who found math and science inaccessible, I could convert these subjects into a language they could understand. Perhaps because I don't sleep with equations and formulas dancing in my head, I could relate to their confusion. This was harder for teachers for whom it had always made sense. I loved it when the gleam of recognition came into a student's eyes; I knew that feeling of accomplishment and relief all too well.

Yet tutoring also engendered frustration. I didn't realize how much easier it was to teach an individual than an entire class of different abilities and needs. My naivete, however, was probably a blessing. It led to my quixotic belief I could be a teacher as well as a tutor. I would do it differently. I'd be creative and patient, and I'd reach those students for whom math and science didn't come easily. I ached to spark young minds to see beyond the text, to use science to discover the natural world. My hubris was boundless, until that first day when those seventeen pairs of eyes drilled through to my truth: I had no idea what I was doing.

---

Just as I had fallen into the Bald Eagle project by accident, I fell into my first teaching job in the same way. I answered an ad for a part-time math teacher position at a local private K through 9 school. With no teaching credentials and no certification, public school was not an option. Seventh- and eighth-grade math sounded doable, and a part-time job could introduce me to the world of teaching without consuming me. My son Michael was only in first grade, and I was hesitant to commit myself to anything more.

When the head of school held up my resume at my interview and asked, "It seems you have more biology in your background than math. Would you be able to teach that too?" I was elated. The possibility of teaching biology hadn't occurred to me. I spent the summer rereading all my math and science textbooks and wondering how on earth I was going to pull off this feat. It felt like 1976 all over again.

---

During my first year, I made many assumptions, almost all of them wrong. In my fantasies, all students were bright, enthusiastic about learning, and ready to absorb every morsel of enlightenment I could bring them. There was no room here for the scowling boy who slouched in the corner and never did his homework, the posse of girls who whispered throughout the class period, or the parents who became irate when their child genius received a B–.

At my one-month review, the headmaster asked me how things were going and if I had any issues. "Just one," I said. "Discipline." He recommended I adopt a sterner approach to pedagogy. Not my style. I learned quickly that the students weren't my issue; I was. How could I infect slouchy boy with enthusiasm for learning if I couldn't let go of the rule book? Once I realized that I needed to replace discipline with humor and compassion, things improved.

If I could meet students where they were, I could engage them by making biology relevant. In those years, bio textbooks began with the cell, which, in my opinion, was the most onerous aspect of natural science for fledgling scientists. From there, with total confusion rooted in mitochondria and Golgi bodies, we were to move onto cellular respiration and photosynthesis. All was invisible, not to mention incomprehensible, even under the most powerful microscope (which we didn't have). These processes were left to the fourteen-year-old imagination, keyed at that age by hormones to many more salient topics.

I decided to reverse the textbook, starting with the last chapter: Ecology. This was the one that we never even reached when I was in biology class yet was the most important and relatable. Instead of labs with Bunsen burners and petri dishes, we went outdoors.

---

Twenty years after my first day teaching biology, I was no longer a neophyte. Instead of opening with the dreaded textbook, my first lesson began with a poster. Penguins on top of an iceberg, hurtling themselves into the sea below, one at a time. Lined up as if patiently waiting for the opera house to open, they resembled short men in evening dress. I told the class to study the image and decide what was happening and why. The range of responses was revealing. *The first falling penguin had been pushed by the ones behind; they were hungry and the hungriest dived first; they were committing suicide; there was a predator lurking in the sea below so they were waiting to see if the first diver would be eaten.* All good answers that made perfect sense if you knew nothing of the colonial behavior of penguins. I would pull out the poster again in May to see what they had learned about natural selection, genetics, predator-prey relationships, and of course, all those unpronounceable words that explained such phenomena. What they discovered that first day of class is that there are few right answers, but many right questions. They needed to hypothesize, to take risks, to defend their ideas. If that was all they learned, I had done my job.

I found that the more involved the students were in their own learning, the more likely they would be engaged. As the poet Anne Sexton once told herself, "Whatever you do, don't be boring." Starting with ecology was one of the best ways to accomplish this. Learning about the world around them was a perfect lead-in to understanding the more complex concepts of genetics, evolution and from there, the cell and its workings. Not the other way around.

We spent the first semester discussing the interactions of living beings with their ecosystems, ending with the ultimate drama: human impact on the earth. The final project of the term was a debate on a current controversial environmental topic. Oil drilling in the Alaskan National Wildlife refuge, wolf reintroductions in Yellowstone, construction of wind turbines off the New England coast, elephant culling in Kenya—all in the news at the time with two definite sides. Each student was given a position and role to play—the CEO of Exxon-Mobil, the director of the World Wildlife Fund, a Kenyan farmer, and so on. Sometimes their passion got the best of them.

One year, I was summoned to my supervisor's office for a "talk." Apparently, the ninth graders had carried their debate on deer hunting in national wildlife refuges into the cafeteria.

In both the pro and con camps, the argument had boiled down to whether the deer should be left alone to overpopulate and risk starvation over the winter or whether it was more merciful to cull the herd with professional shooters and donate the meat to the hungry. A moral dilemma with no right or wrong answer. This situation taught me how difficult some environmental issues are to mitigate, especially when no solution will be ideal.

---

Although teaching ecology was rewarding in that it was relevant to the students, even those for whom science was anathema, I knew what lay ahead. Just as they were starting to relax and accept the rigors of science, were earning high marks in a new language—with recognizable words like *populations*, *communities*, *species*, *ecosystems*—I had to raise the bar. Skipping backwards to the front of the book, we left the global ideas behind us. Entering the microsphere, I eased them across the boulder-strewn path towards cellular processes and biochemistry, all topics they needed to master.

Drawing from my own serpentine path to science, reliant on discovering the natural world outside the lab. I wanted to reproduce my college experience of the bird walk, that life-changing moment when wildlife became visible to me. I needed to build a nature trail.

The campus was made up of buildings, playing fields, and parking lot; the remainder of the nearly forty acres was an impenetrable red maple forest with a dense understory interspersed with wetlands. After convincing the school administration to let me build a two-mile perimeter trail around the school property, I hired a landscaper to help me navigate through wetlands and around trees, carting chain saws and clippers, following my neighbor Ralph's lead at our farm. The new trail allowed students access to the woods and encouraged teachers in other disciplines to incorporate nature study into their curriculum.

My hidden agenda, however, reached beyond the educational value of the trail. I had provided a wildlife sanctuary for the neighborhood. Between the farm and school, I was fulfilling my dream to restore wildlife through habitat protection. Deer, raccoons, foxes, possums, hawks, turkeys, and other animals began to use the trail, build nests and burrows in the woods, and seek safe havens from the suburban development around the school grounds. Students could now experience the magic of nature during their school day.

The most rewarding parts of teaching were field trips. At the New England Aquarium in Boston, the students didn't just peer through the glass at the Giant Ocean Tank or watch the seals and penguins cavort in their pools but met the researchers studying right whale behavior and propagation of lobsters. Although I was far more excited than the students, it was still good for them to see science in action, to witness both the compelling and mundane aspects of a career in field research.

Overcome by the intensity of the lab, with researchers glued to their microscopes or computers, the class became mute. Scientists are rarely exposed to kids, much less teenagers. A silent standoff ensued, each counting the minutes until the visit was over. The only whisper of enthusiasm came when the right whale biologist related the story of the Rottweiler assigned to track whales by standing in the bow of the boat and pointing his nose toward the floating masses of poop delivered by the leviathans invisible beneath the surface. Poop, scat, excrement, doo-doo—always attention getters with any age student. Once the tension was eased, the scientists and students exchanged ideas, and many came away inspired, bombarding me with questions on the bus ride home.

Whenever possible, I brought wildlife science into the classroom. The students needed to see the objects of the research, not just the scientists performing it. I asked outside speakers to come share their stories and if possible, their animals.

The most popular presenters were the owners of Eyes on Owls, an educational and rehabilitation facility. Mark and Marcia Wilson visited us every year with up to ten species of owls. Students were able to touch the birds, learn what makes owls different from all other raptors, and even were taught to hoot! Seeing the owls flap their wings and clack their beaks at Marcia and Mark transported me back to my eagles expressing their indignation at being banded or rescued from the marsh.

I wanted the students to feel they could make a difference and not just absorb information about animals. We became involved with Grassroots Wildlife (now part of Zoo New England), a program started by herpetologist Bryan Windmiller to protect threatened freshwater turtles. Bryan and his team removed baby turtles from nests where proximity to roads and other hazards spelled their doom. Classes from area schools raised the babies in an aquarium from September to May, releasing them into a local wildlife refuge, well-fed and strong enough by then to manage for themselves and restore their populations. It was like hacking eagles with one exception: the students could handle the turtles and not risk imprinting since turtles never provide maternal care. Not only did the class learn about reptiles but also became more aware of the critical and fragile nature of wetland habitats.

Each time he came to speak to the class about turtles, Bryan would ask if they were aware of my eagle work. Knowing my hilltop eagle tower was a world far removed from that of the middle-aged science teacher who stood before them, I never talked about my previous life. In their minds, I was the keeper of the grade book. But somehow Bryan, as raconteur of my story, invoked their awe and recognition. After his visit, the class would follow up by asking me more about my experience, creating an opportunity to introduce them to the wonder of the Bald Eagle.

My favorite field trip destination was our farm. Our study of ecological succession and creation of wildlife habitat meant nothing when confined to the classroom, but when the students came to the farm, they could see it all in real time. I showed them why we waited until late July to mow fields so the birds could finish nesting; why we managed the invasive plant species so native wildflowers and trees could survive; why we provided brush piles so rabbits, foxes, and fishers could make burrows and dens; why we encouraged bats to live in the barn; why we put up roosting bars in the fields so hawks could watch for prey.

If those seventeen pairs of eyes looked at me suspiciously on that first day of class all those years ago, on these farm trips—especially when I unearthed a vole burrow or caught a garter snake for them to observe up close—they opened wide with incredulity. How could this placid teacher, who seems normal in school, be so at one with the wild world of creatures?

My most revelatory moments came when I introduced the class to the donkeys. Perhaps there would be one student who was willing to feed the gentle pets a treat, but most stood far back from the fence, terrified that old Anthony, at the age of forty-five, might leap out at them. Even as his soft grey muzzle took the Fig Newton from my palm, they still held back. He had to win their confidence, just as I had had to do, and it would take more than one field trip or one interaction. But he did elicit their curiosity, a start. I recalled my fear of snakes dissipating with the touch of a boa's body under my fingers. I knew that these students only needed time and experience to connect to the unfamiliar world of nature, and they too could become captivated.

# Part Five

# Eagles Aloft

## — Chapter 26 —

# Proof of Success

My life after 1978 followed a trajectory that no longer intersected with the Bald Eagle. Once I had left the refuge hillside, the whereabouts of my seven birds, and those following them at Montezuma, was a mystery continually haunting me. I often wondered how they had fared in the wild, but sadly it is only when an eagle carcass is recovered, and its band identified, that we can know its fate. Color tags and radios drop off after a few years, surveillance becoming impossible. By the late 1980s, whenever an occasional Bald Eagle was spotted flying over Massachusetts roadways, along rivers, or perched in trees, I would fantasize that it might have been my bird. Perhaps one of the seven had migrated east from New York and was on territory nearby? Convincing myself that it was survival of the species that mattered rather than the individual brought me solace when I became obsessed with curiosity. I tracked the activities of the hacking programs in New York and Massachusetts whenever I could, even though it made me nostalgic for my days on Clark's Ridge.

During Cornell's two-year oversight of the Bald Eagle project, my role had been to work out the logistics of the initial hacking experiment. Future efforts focused on restoring the New York population of our national bird. This objective depended on how well Bald Eagles could adapt, without parents feeding and protecting them, to survive and reproduce.

My seven birds had thrived until it was time for them to leave the refuge in the fall. Some were known to be feeding themselves, some even returned to the refuge confirming their longer-term survival skills, and all had given us valuable data needed to continue the program.

My research offered a lens into the feeding, fledging, and intraspecific behavior of young eagles.

When it came time to hand over my thesis to DEC, so they could continue the program in my absence, I wrestled with mixed emotions even though I knew my job was done. The program and its young subjects had been so much a part of me it was difficult to pass it on. It was time, however, for others to take up the reintroduction effort and carry it to completion.

DEC decided to carry on the Montezuma project with few changes for the next four years to monitor the eagles' responses to their new environment. About four eaglets were released each summer from the Clark's Ridge tower between 1978 and 1981, thanks to donations of young birds from the Great Lakes and Patuxent.

While Montezuma continued to be the premier hacking site for a while, everyone in DEC was aware that additional release sites needed to be found. If eagle behavior was true to form, the birds would return to their natal area, limiting the numbers of pairs that could nest in the refuge. After twenty-three eagles had been released into Montezuma by 1981, it was believed the refuge had reached carrying capacity. It was time for changes to be made.

DEC's long-range plan was to increase New York's eagle population to ten breeding pairs. In addition to locating alternative release sites, spurred on by their wish to put as many eagles as possible into the sky, biologists needed to find additional sources for these birds.

The Great Lakes eagle population wasn't large enough to meet New York's increasing demands. Patuxent also couldn't supply enough captive-bred young to meet the new quotas. The science of artificial insemination of Bald Eagles was on the horizon, but still in its infancy. Stanley Wiemeyer and his team remained dependent on the unpredictable nature of mating, courtship, and breeding of adult eagles in captivity.

As a result, after 1981, nests in Alaska became the obvious choices as a source of eaglets for New York. The Alaskan population was healthy and unaffected by chemical pollutants. When we were considering which states could provide Bald Eagles for the first reintroduction in 1976, Alaska had been ruled out as a possibility because of concern the subspecies there might be too different genetically from the northeastern population. After the releases proved successful, however, this decision to use Alaskan nests was reexamined.

---

As it turned out, there was plenty of room at Montezuma for multiple pairs. Even though I was no longer involved in the releases, I kept my eyes and ears out for positive news reports about any hacked eagles. I was thrilled to learn about the successful "Trio," a family unit of eagles consisting of two males and one female who cooperated to raise young. The three birds built multiple nests successfully in and around the Storage Pool. One of the males had been released in Montezuma in 1978, and the other male from the Oak Orchard Wildlife Management Area, a later release site midway between Rochester and Buffalo. Both could be identified as hacked birds by their yellow tags. The female wore no tag so was not presumed to be a reintroduced eagle.

By 1994, there were two eagle nests in Montezuma—the Trio's and that of another pair who also built a nest in the Storage Pool. By 2021, Montezuma reported a grand total of six successful Bald Eagle nests, proof that our choice of site for the initial hacking program in 1976 had been a good one. The refuge, originally chosen because of the food and protection it could offer the birds, has come to be known as a major nesting and wintering area of Bald Eagles in New York.

As the years passed, my connections to Cornell and Montezuma faded. In 1984, Tom Cade moved The Peregrine Fund to Boise, Idaho, where it became part of the World Center of Birds of Prey. The staff at Montezuma changed over many times during the years after the hacking project, so I no longer received reports of eagle activity. I was surprised, therefore, when a friend, driving east from Ohio, sent me a photo of a massive Bald Eagle sculpture on Montezuma refuge property off the New York Thruway. It seemed the refuge was embracing its eagle population in a visible way.

In 2016, local artist Jay Seaman was commissioned by Tom Jasikoff, the Montezuma manager at the time, to create the statue on the perimeter of the refuge bordering the Thruway. Measuring twenty-two feet wing tip to wing tip and weighing 1,300 pounds, the work was done to commemorate the fortieth anniversary of the first Bald Eagle release program in the refuge. I was glad that the success of 1976 was being recognized in this dramatic way although I was surprised that Cornell hadn't been involved in the celebration. As time passes, much is forgotten.

Peter Nye, who was sent to Montezuma to learn the hacking techniques while I was there, took over the DEC program after my departure in 1978. Pete continued to oversee all Bald Eagle reintroductions, in addition to other species, as the director of New York's Endangered Species Unit until he retired in 2010. With the help of Mike Allen, a valuable member of the crew on the hilltop in 1977, these two men were instrumental in helping restore the Bald Eagle population in New York. Sadly, Mike passed away in 2017.

By 1982, DEC created four other hack sites and released large groups of 25 to 30 birds per site, reaching a total of 198, until the program ended in 1988, once New York's original goal of 10 breeding pairs had been achieved. By then there were enough eagles in the sky to encourage breeding and dispersal throughout New York and into the neighboring states. The number of eagle breeding pairs in New York by 2020 were estimated at more than 400.

Other states soon adopted the hacking technique and initiated their own reintroduction programs.* As of 2015, there were more than one thousand eagles hacked in the eastern United States, with Tennessee the last state to reintroduce eagles by hacking in 2016.

Besides the donations from Alaska and the Great Lakes states, eaglets for release were being provided by the Canadian provinces of Manitoba, Saskatchewan, and Nova Scotia. The genetic diversity of the species has increased enormously as a result. Once a concern, many biologists now consider this a bonus, for an enlarged gene pool usually bodes well for a

---

* Massachusetts and Vermont in New England; Pennsylvania, New Jersey, and Washington, DC, in the mid-Atlantic region; and North Carolina, Tennessee, Georgia, Kentucky, and Alabama, in the South.

species' ability to adapt and survive. As the eagle population continues to rise, their behavior might be evolving to make them more successful. Bald Eagles have always been more wary of humans than Ospreys, more prone to abandon a nest site if human disturbance comes too close for comfort. In recent years, however, observers are finding that the birds are becoming more tolerant of human proximity and noise.

---

The renowned success of the hacking program has raised the obvious question: How many are too many? From a mere 417 pairs in 1963 in the lower forty-eight states, there are now more than 71,400 nesting pairs (according to the 2021 USFWS report). In Maine alone, there are between 780 and 915 active nests. People are enthralled by the sight of a Bald Eagle flying overhead, but the population density of such a large predator has its drawbacks as well.

In Alaska, eagles prey on seabirds, many species of which have their own issues. Bald Eagles are even sometimes called "trash birds" because their scavenging behavior leads them to feed on garbage. The Great Cormorant is disappearing from Maine's offshore islands largely due to eagle predation. The beloved Osprey is vulnerable to the larger eagle, for the Osprey's reliance on fish for food requires the parents to leave the chicks unattended in the nest long enough for a Bald Eagle to snatch them. Some Osprey lovers, who erect nesting platforms to bolster their birds' populations, can be vocal critics when a Bald Eagle sabotages their efforts. Although they don't tend to nest on Osprey platforms, Bald Eagles have been seen using them in winter as hunting perches and then hanging around to prey on the adult or young Ospreys once the nest becomes active in spring. In fairness to the eagles, there have also been reports of Osprey adults attacking eagles whom they see as rivals for food, so the interspecific competition can sometimes go both ways.

As with any predator reintroduction, there are often unforeseen and unintended consequences as the food chain is impacted. For most people, however, the tendency of a Bald Eagle to occasionally take a popular prey item is a small price to pay to have the majestic white-headed bird back in our skies once again. There is nothing that compares with its grace, its power, and the freedom that it has come to symbolize. Bringing the Bald Eagle back from the brink of extinction is a conservation success story that hopefully can inspire future efforts to save species that need our help.

When referring to the quadrupling of Bald Eagle numbers between 2009 and 2021, Deb Haaland, Secretary of the Interior and the first Native American cabinet secretary, helped me see how my experiences with eagles and the Iroquois became so intertwined: "The Bald Eagle has always been considered a sacred species to American Indian people. Similarly, it's sacred to our nation as America's national symbol. The strong return of this treasured bird reminds us of our nation's shared resilience."*

* "'A Historic Conservation Success Story': Secretary of the Interior Deb Haaland," Indianz.com, March 24, 2021, https://indianz.com/News/2021/03/24/a-historic-conservation-success-story-secretary-of-the-interior-deb-haaland.

## — Chapter 27 —

# A Special Eagle

When I said goodbye to my eagles in the falls of 1976 and 1977, my inability to guarantee their future shrouded any relief I might have felt at setting them free. Many threats were facing them as they made their way out of the refuge area. Random shootings, roadside hits by automobiles, and encounters with trains while eating carrion on the tracks, are among the most common causes of death of eagles, and none of these can be foreseen. I had reintroduced seven birds into the wild, but the question remained: How many would survive to reproduce?

Because Bald Eagles don't reach sexual maturity until about three years of age, I had to wait to see if any of them would succeed. But I didn't have to wait long. Tom Cade, aware of what the news would mean to me, sent along an astounding update in 1980.

In Watertown, New York, eighty-four miles north of Montezuma, a Bald Eagle pair had built a successful nest, the first seen in the state since 1973. The only other remaining nest, at Hemlock Lake, was inactive, its resident eagles not having produced any young for seven years.

The report of the new active nest was dramatic enough, but to add to everyone's surprise, the pair was W1 and W2, the Wisconsin nestmates I had released in 1976! This duo had gone their own way for several years but then reconvened when they were ready to breed. Whether it was for lack of other mate options or whether their bond had transcended their close kinship, they had chosen each other in the end. They had two young that year and fledged one eaglet—the first to be born as a result of the reintroduction project. The news of their pairing elated everyone involved, for here was proof that the hacking technique had produced eagles who could not only survive in the wild, but who could breed and successfully hatch young.

I was overwhelmed at the news of the mating of W1 and W2, but not as surprised as others might have been. The bond between these two Wisconsin nest mates had been strong from the beginning, the two eaglets never being far apart even when they were in the marsh learning to hunt or had left the refuge to explore the environs. I saw them leave at the same time on the first day of hunting season that fall and had held on to the hope they would stay together. W2's return to the refuge alone in 1977 and 1978 gave me pause, so I was thrilled that they had reunited. Of course, there was a practical explanation for their reunion. Aside from the Hemlock pair, they were the only eagles in New York State of breeding age at that time, the other hacked birds being too young. It only made sense they would find each other when their instincts convinced them it was time to build a nest.

---

The next year brought more astounding news. The Hemlock Lake nesting pair had not had any luck producing young, a fact scientists attributed to the female being infertile or too riddled with chemicals to produce viable eggs. But DEC biologists wanted to take advantage of this pair of eagles who seemed so loyal to their nest site, and to each other. Led by Peter Nye and Mike Allen, a fostering program was initiated whereby captive-bred eaglets from Patuxent Wildlife Research Center* were placed in the nest for the pair to raise as their own. The Hemlock female, who hadn't lost her maternal instinct despite her infertility, successfully raised eight eaglets this way from 1978 to 1983.

However, her foster mother reign was curtailed by the loss of her mate in 1981 when he was found shot to death nearby. All was assumed lost until one day that spring when she reappeared with another male eagle. She had found a second mate and had resumed her position at

* In 2020, Patuxent in Laurel, Maryland, and Leetown Science Center merged to create the Eastern Science Center (EESC).

the nest site, eager to receive more foster eaglets. You have to admire her determination to raise young against all the odds.

The female was thought to be quite old for an eagle and when she disappeared in 1983, it was assumed she died of old age. The male, not to be deterred and apparently attached to the area, found another mate and continued on at the nest. This time his new female was fertile, and the pair raised their own young for the incredible span of thirty-six years, producing a total of about seventy Bald Eagles. This male, who had adopted the Hemlock Lake nest, was no other than my rehabbed eagle M3, the most complacent of all the eagles I had released, the one who never gave me any trouble, and who didn't even require the blue-jean sleeve, so calm was he during our rescue forays out of the marsh. Registered by his band number 03142, he had survived on his own, found a female to pair with, a nest to adopt, and gone on to find a second mate to become the most productive and longest living Bald Eagle in known history. My conflicting feelings of joy and sadness upon reading the press release about M3's recovery by the side of the road in 2015 resulted from my own parental pride in his accomplishments.

I have often wondered if the eagle bander who discovered M3 hunkered down in a nest on the edge of Puposky Lake, in the Ottawa National Forest north of Bemidji, Minnesota, his leg broken so he couldn't stand up, knew what a life this bird would come to lead.

Treated by strangers at the University of Minnesota Veterinary College and then shipped halfway across the country to a hillside on the edge of a marsh, the eaglet was plopped into an unfamiliar nest with unfamiliar eagles. He was never to see his parents or siblings again but somehow was expected to form an attachment to this area of New York so he could help restore his species. We asked much of this eagle, and he gave us even more. Not only did he mate with the female of the only existing nest in the state, but he held fast to that nest site until he could find a new female to replace the first one after her death.

M3's loyalty to his mate and breeding grounds seems unusual but then again, we know little about the strength of Bald Eagles' bonds. It is even hard to know how many young eagles were fledged from that nest who could lay claim to M3 as their father. Until he was found on the side of the road and identified by his band, nobody even knew he was the male eagle in the Hemlock Lake nest. If he and his second mate were able to produce an average of two eggs a year, hatch both, and see the young survive to breeding age, he can be credited with many of the eagles we see in the skies of New York and the surrounding states today.

I had known he was a special eagle upon first meeting him back in 1977, when I transferred him from his crate into a man-made nest of strange birds, hoping he would be accepted by his peers. He was a survivor even then.

---

Of the seven eagles I had released, I knew three (W1, W2, and M3) had survived to reproduce. I never found out the whereabouts or fate of the other four birds. From 1976 to 1981, out of the twenty-three eagles released from Montezuma, thanks to the combined effort of Cornell and DEC, my three mated successfully, one 1978 release became part of the Trio nesting group, three were found dead, and the fate of the other sixteen remained unknown.

Wild eagles are thought to have a 50 percent survival rate in their first year and a 20 to 30 percent chance beyond that. Although hacking eagles couldn't increase their chances of avoiding all the threats that impact these birds in the wild, it didn't appear to make them worse, a realization that boded well for future conservation efforts.

## — Chapter 28 —

# Coming Home: Montezuma 2022

*Rafts of wintering ducks and Canada geese rest on the river's flotsam. Barely visible against late autumn's platinum sky flies a dark bird with wings outstretched in a perfect plane. Finger-like wingtips, the absence of flapping, and the nearly seven-foot wingspan stop my breath as I recognize the adult eagle bearing down on the flock, then veering upwards with jet-like precision at the last moment, heading towards the woods behind where we are standing. Brave souls, these waterfowl, never flinching. How do they know the raptor is playing with them and not hungry for a duck dinner? A common sight for the other observers standing with me, fortunate to be out here every day. The only gasp comes from me, one of many I will utter during this visit.*

Several more eagles fly over my head toward a group of red maples hugging the inner tributary. Adults land on tree limbs overhanging the dike road, while the immatures, six altogether, with their uniforms of chocolate brown from head to toe, perch like preschoolers in one of two adjacent trees. I want to call the class to order, to watch

Adult pair perch together on a snag near their nest site in 2022. *Photo courtesy of Jackie Bakker*

and listen and tell them where they have come from and what it means for them to be here on this day. I want to tell them about their grandparents, those stalwart seven eagles and the others who followed them, whose resilience and persistence paved the way for their return to this refuge nearly fifty years ago. I want to stand there, alone with my birds, until they can grasp their own significance.

Immature eaglet perching near marsh in October 2021. *Photo courtesy of Jackie Bakker*

I have returned to a planet, long abandoned, not quite foreign yet existing in shadow, marked by cloudy images tapping gently on my memory. The red maple snags, where my eaglets once perched to learn about the world around them, no longer dot the marsh. Only cattails remain, and even these have been beheaded by wind or water. The once tall grasses form endless greenish-brown clumps beyond the open channel, reaching to where a hillside begins to rise above the marsh.

The water is high, the current swift. It would be impossible to cross to the field of cattails in the distance without a boat. It would be at least chest high and require waders to navigate it.

But I know it is possible. In this mass of vegetation, interspersed with standing pools of water invisible from the outside world, had once been seven hapless eaglets, each rescued from certain disaster by a young researcher, so desperate to reach her quarry she didn't notice how deep the water was, how thick the grasses were, how many times she'd fall into the cavities below her. *Had that really happened? Surely, the river today must be wider, deeper, the Erie Canal forcing manipulation of the water levels over the years.* As I begin to question my memories, seeking clarification at every step of my mind travel, I look up at the hillside. My hillside. Nothing but trees can be seen from down here on the dike road, but wait—there is a shadow behind the tallest ones, a bare silhouette of telephone poles, straighter and darker than trees. Something is there but I can't be sure.

---

Many people love to tell me about the Bald Eagles they have seen. Most don't know I had been involved with eagles but are astonished by how many are appearing in their backyards, flying across highways, perching in trees near their lake or ocean front homes. Occasionally, I will say, "I was part of a program to bring eagles back to the Northeast, so I am glad to hear this." But usually nobody takes this seriously or just doesn't understand they had ever been gone. We have learned the lessons of extinction from the Passenger Pigeon, the Dodo, and the Great Auk, but few recognize how close the Bald Eagle had come to being one of them. I knew people needed to be reminded, the eagle's story needed to be told.

To do this, I had to complete the circle. It wasn't enough to know that Bald Eagles were repopulating our region in great numbers. I had to go back to where they had begun. I wasn't ready before this to return to Montezuma. My two years there had been such a formative part of my life that I dreaded disillusion. Whatever had kept me away was now replaced by a hunger to revisit my hillside, my marsh, and most of all, to see my eagles again, or at least their progeny.

In November 2022, I took a trip to Cornell and Montezuma to settle my mind. I had forgotten many details of the geography of the refuge and needed confirmation that things were where they were supposed to be.

I had read about the Trio—the first hacked eagles to nest in the marsh—and the statue out on the New York Thruway. I had seen bird lists mentioning Bald Eagles in the refuge and along the lakeshore in winter. But my belief was suspended, so inconceivable was the idea that the eagles had returned. I needed the tangible, to see the birds soaring above me, to view their nests crammed into the Vs of the trees, to witness the adults with their white heads and tails circling over me. Only then could I know for sure that the project—the dream, really—of 1976 had been true. And I had to see this before fifty years had passed.

Larry offered to go with me, partly to keep me company and serve as what he termed my "research assistant" and photographer, but also to revisit his own memories. Ithaca had been a special place for both of us—where we had met and discovered our shared passion for the environment, where we had learned how to make our contribution to a better earth. He had a long list of friends and colleagues he wanted to see; I had the eagles. The Peregrine team was long gone from Cornell: Dr. Cade had moved to Idaho and had passed away in 2019; Jim, Phyllis, and Willard had relocated; my professors had retired and left Ithaca. I was relieved, in a way, to be able to focus on the refuge and the research I needed to do.

It felt strange to call the Visitor's Center and introduce myself, knowing the staff hadn't been there in 1976 and would have no idea who I was. A stream of gibberish flowed from my mouth as I tried to explain my role in the eagle reintroduction more than forty years earlier, and why I wanted to return to the refuge. Each person I spoke to was surprised to learn there was a connection between Cornell and the hacking program. Certainly, nobody had any idea I had been involved. However, the pieces began to click into place. The disembodied voices became more focused as their surprise morphed into curiosity, then into urgency to fill in the gaps of a history they only knew part of. At last, they were receptive to, and then even enthusiastic about my visit.

I was surprised how much more extensive hunting had become since the 1970s. I had to plan the trip around the deer and waterfowl seasons now extending from September to January. Larry and I planned to spend one day at Montezuma and the rest of the visit at Cornell, retracing our steps through time.

Sleep evaded me the night before we left for the refuge. My nerves jangled as they always do before any new encounter. Fear of not knowing how to act or what to say, the anticipation of what might happen, made it impossible for me to relax my mind.

When we finally got on 89 heading towards Montezuma, up the west side of Cayuga Lake, I was struck with the unfamiliarity of it all. I must have driven this route dozens of times in the two years I was in the refuge, but nothing seemed recognizable. Today, 89 is dotted with vineyards, marked by fancy wooden signs and arrows pointing toward the lake or up the hill. I don't remember vineyards. I don't remember huge lakeside houses with boathouses and docks and sprawling decks over the water. Was I so intent on getting to my destination, whether it be back to the eagles or home to my apartment, that I failed to notice my surroundings back then? Or can a forty-seven-year time warp be this dramatic?

I wondered what to expect when I arrived at the refuge. *Would it be an anticlimax, a disappointment, an alien place?* A giant welcome sign at the entrance was adorned with an image of a regal Bald Eagle taking center stage—the first indication that things were different. The refuge headquarters building was not where I had remembered it, but how many times had I really been there, so tied was I to my hillside back then?

Everyone was waiting for us when we arrived. There were four in the welcoming committee: Bill Stewart, the refuge manager, and Linda Ziemba, the staff biologist, the professional arms of Montezuma's eagle presence; Jackie Bakker and Pete Saracino, dedicated volunteers in charge of bird surveys and public education for the refuge.

Although all four were happy to meet me, it was Jackie, the protector of the refuge's eagles, who confirmed my belief I had done the right thing by coming here. "I can't tell you what it means to me to meet you!" she exclaimed as she rushed breathless into the office after the rest of us had assembled. "You have made my day, maybe even my week!" All my anxiety drifted away when I heard her words and felt her enthusiasm and joy.

Their curiosity about me and my program put me at ease, for here were people as excited about the eagles as I was. Usually when I spoke about my work, I worried I sounded like a stuck record. But these folks were listening to every word, recognizing I was revising the history they had been handed. As I spoke about the program and related the story of those first seven eagles, I looked from face to face, all riveted on this new information pouring from my mouth. In all the records they had, the eagle hacking project had begun when DEC took over the program in 1978. They had no idea Tom Cade and Cornell's Peregrine team designed the project, that it had originated in 1976, or that I even existed. To find out there was more to the story than this was mind-boggling to them. Larry observed afterwards they treated me like a celebrity. From the time I walked into the refuge headquarters until we left in late afternoon, I was given the keys to the Montezuma kingdom.

When I asked if I could go back up to Clark's Ridge, I was told the road had disappeared, that there wasn't even a trail kept clear to the tower. There were only poles left, no platform remaining. I still wanted to go. We piled into the refuge's Tahoe, a sturdy four-wheeler, and pushed through the undergrowth, through the gate *(had there been a gate back then?),* and up the former dirt road, now a path barely wide enough for a vehicle and covered in brambles, the kind that follow you by sticking to every part of your body. It finally became impassable. We walked the rest of the way, a few hundred meters up to the tower, adrenaline pounding through my body as I anticipated what I would see. I was traveling back in time and frightened about the toll so many years had taken on my memories of place.

The ruins of the hacking tower still standing in 2022. *Photo courtesy of Larry Morris*

The poles were there, all eight of them, with the metal brackets still attached. I recalled my more dexterous body scaling those steps several times a day and couldn't imagine it was the same self. My septuagenarian muscles and ligaments twinged with regret and denial returning to that time. The platforms and blind were gone, the empty spaces left to imagination. What surprised me most was the box for the video camera, still dangling from its own pole flanking the tower, its back gone but the housing intact as if the cameras were still rolling, recording this visit just as it had the story of the eagles so many years before.

The most disturbing sight was the forest, a creeping scourge of vegetation threatening to overtake the tower and eventually the whole site. I thought of Tikal, the Mayan temples in Guatemala, where the jungle had absorbed the man-made edifices to the point where nobody even knew they were there. One by one they had to be uncovered, exposed, and then invaded by tourists like me, searching for the past.

I pictured the flight of M2, the smallest of the seven eagles and the last one to leave the nest, who tried to reach a tree on his first attempt, and how far that had been from the platform. Watching him fly for the first time and grab onto a maple branch, only to plummet down into the dreaded cattails below the ridge—the despair I felt all those years ago came flooding back. Now the tree where he had landed was hidden by the advance of fast-growing aspens, little more than thirty meters away. Likewise, the marsh seemed far in the distance, woodlands now covering the slope between the cattails and the tower. It made sense why the tower was nearly invisible from the road below.

I yearned to be alone with the ghosts of Clark's Ridge, to allow my memory to travel backwards, to sift out the details that I knew time would recover but were eluding me in the moment. I wanted to put myself back in the blind, to be sitting near M3, that special long-lived eagle, the embodiment of determination and serenity, and the other six eaglets who were so important to my life. Unknowingly, they had changed the course of my future, had been the symbols of resilience and hope, convincing me of the importance of wildlife conservation and my need to be a part of it.

Instead, this group of enthusiastic and generous people, who were trying to travel back to 1976 with me, were bombarding me with questions, exuding eagerness, searching for clues to the past. *What did it look like? How did you feed them? Where were the platforms?*

For a few minutes, I wandered away from them and stood below the tower, my hand on the pole I had climbed so many times. I could hear the eagles screeching above me in the nest, calling for the fish they knew I would be bringing to them. Soon they would be leaping into the air, wings flapping, until they took off down the hill to the marsh, gliding smoothly into the cattails, oblivious to the dangers awaiting them on this first day of freedom. How could I explain these images to the group standing behind me?

Eventually, it was time to look for nests. Jackie Bakker, in charge of monitoring each site, was eager to show me the six active ones. As we drove along the dike road, which remained closed to the public, I asked to see the spillway where I had once caught carp and built my holding cage in 1977.

Each nest sighting left me without words. Everyone must have thought I was partly mute during this tour, for I couldn't express what was going on inside my brain. *How could this have happened? How could these eagles have been so willing to do our bidding, to return and breed where their grandparents had once been raised by humans and then released in a project that by all counts had no right to be so successful?* Resiliency is the only word. Yet the nests, once alive with young eaglets but left

Adult eagle guarding its one nestling in nest in May 2020. *Photo courtesy of Jackie Bakker*

abandoned until the next breeding season, were inanimate, empty vessels fueled by my imagination on that cool November day as Jackie told stories of each pair; she knew them all so intimately. It was the eagles themselves, flying overhead, perching near but never on their vacant nests, who reached into the depths of my being. They had accomplished the unthinkable; they had come home.

---

Besides the changes to Montezuma over time—the expansion of its boundaries, the growth of vegetation, the blowdown of trees in the marsh, the higher water levels—I was stunned by the position the Bald Eagle had earned as the iconic brand of the refuge. The bird had indeed adopted these marshes as home. Besides the six active nests, producing eleven eaglets in 2021 and seven in 2022, the wintering population of eagles confirms their choice of the refuge as prime habitat and feeding grounds. In January 2018, seventy-eight eagles were seen in one day in and around the Storage Pool. Often the nesting pairs build two nests, using only one but keeping the second in reserve in case they lose their

The author and refuge volunteer Peter Saracino pose with Bald Eagle statue. *Photo courtesy of Larry Morris*

nest tree in a storm. I had to laugh when I heard this, imagining the eagles, whose existence had once been precarious and of deep concern, being so successful they could now afford a second home.

I had been prepared for the statue out on the Thruway, had seen pictures of it online, but its size and magnificence, bronze head and tail gleaming in the sunshine, filled me with reverence. Aside from the statue, images of the eagle followed us everywhere—on the entrance sign, in the refuge headquarters in the form of a bench with a carved eagle at either end, in portraits hanging on the office walls. An elaborate eagle display in the Visitors Center captivates all who visit Montezuma, telling the story of the bird's allegiance to the refuge.

Before our return to Ithaca, Jackie and Pete insisted on taking us to lunch in Seneca Falls. The thought of revisiting the town where showers, laundry, and cheese Danish had helped elevate the comfort of my sojourn at Montezuma was too enticing to pass up. I had visions of reliving more memories once I found the apartment and supermarket,

Adult eagle watching for fish from a tree stump in Tschache Pool in 2023.
*Photo courtesy of Jackie Bakker*

where I had escaped the confines of my tent. Instead, too much time had passed, altering the town until it was a foreign place, not the familiar oasis I expected. Stores and restaurants now lined the main street, manicured parks adorned former vacant lots, the supermarket had moved out of town. Like the refuge itself, Seneca Falls had ensnared me in a time warp, forcing me to doubt my memories. Jackie and Pete treated us to a companionable lunch at a restaurant I had never seen before. On our way out, I asked the waitress how long it had been there—about thirty years—to assure myself it was my recall and not the town that was cloudy.

Pete, a Seneca Falls native, then donned his tour guide hat. He regaled us with the story of the women's rights movement and its origins at the Seneca Falls Convention in 1848. Led by abolitionists and suffragettes Elizabeth Cady Stanton and Lucretia Mott, this was the first time people had gathered to fight for the right of women to vote. Even though it wasn't until 1917, when the Constitution was amended, that women secured that right, the fight had begun here. The Wesleyan Methodist chapel, site of the convention, has been renovated and is now part of the Women's Rights National Historic Park opened by the National Park Service in 1982. When I heard this, I silently wished it had been there in 1976. Perhaps with a physical reminder of the importance of the convention, I would have felt the visceral connection between my work with the eagles and the role of females in raptor conservation. As it was, I was too preoccupied with the here and now, making sure the eagles survived, to allow myself to view the bigger picture. This stop, combined with Pete's interpretation of the town's illustrious history, was a vital part of my return to the refuge. My work here had been bigger than seven eagles. It had supported a legacy of women's activism and a movement whose reliance on perseverance, patience, and courage had changed the country for the better.

After bidding goodbye to Pete and Jackie, Larry and I headed back down the east side of Cayuga Lake to Ithaca. My mind was whirling with all I had seen and heard during this day; it would take months to process the events of the past hours. But I can now be certain of one truth: the eagle has made its way back, not only to these marshes and adjacent woodlands, but into the hearts of the people who come to enjoy its presence. The Bald Eagle has become the refuge's main attraction, a far cry from 1976 when its appearance within the boundaries was top secret, hidden from everyone but us. From here, the seven eagles and their successors have branched out to other states, other regions, to build nests, to raise young, to dazzle the minds of younger generations, until their disappearance has become a distant memory, but one that should never be forgotten.

# Epilogue

Finches and sparrows, nuthatches and titmice flit from feeder to feeder, garnering one seed at a time in tiny conical beaks, only to return minutes later for more, never satiated due to the snow hiding their natural larder. My grandsons, aged two and three, wide-eyed, stand transfixed at the window. Soon the mixed flock of blackbirds will descend and wipe out the feeders, leaving chaff and shells in their wake. Or perhaps the grey and red squirrels will do battle for feeder control. Or the resident Cooper's Hawk, who whiles away its time nearby, out of sight until one straggler fails to pay attention, will swoop into the bush, emerging with grey or brown feathers in its talons. We watch the play unfold from the hide that is our porch as I think of another time, of another meal, of another group of unsuspecting young who lay exposed to outside threats. My ability to be an observer today instead of a protector is not lost on me. My desire to share this avian feast with the next generation reaches far beyond generosity of spirit. I need the children to understand.

When Rachel Carson, in 1962, connected our postwar use of DDT to the decline of songbirds, skeptics surfaced to denounce her findings. As Linda Lear, Carson's biographer, later wrote in her introduction to *Silent Spring*, "In postwar America, science was God, and science was male."* When other scientists explored the long-term effects of this pesticide in birds of prey, explaining their precipitous decline, the public finally began to take notice. Less than two years after her book's publication, Rachel Carson was dead, a fifty-six-year-old victim of metastatic

* Rachel L. Carson, *Silent Spring* (New York: Mariner Books, 2002), xi.

breast cancer; but her words lived on. Carson knew that change, if it happened, would be slow.

Yet six years after her death, the first Earth Day was created on April 22, 1970, spearheaded by Senator Gaylord Nelson of Wisconsin, in response to rampant air pollution and a major oil spill off the Santa Barbara coast. The environmental community was galvanized into action. The Environmental Protection Agency (EPA) was established and charged with monitoring air and water quality in December of that year. The Endangered Species Act passed three years later, listing the Bald Eagle. People were listening. Just not carefully enough.

DDT was banned in this country in 1972, but exports of it continued for several more years. Today, it is still manufactured in India (and possibly North Korea), and is used in countries in South America, Africa, and Asia to combat insect-caused diseases like malaria. Many of these countries are the wintering grounds of birds of prey, warblers, and shorebirds that breed in the United States and Canada. There was a ceasefire in the war to save birds from persistent chemicals, but not an armistice. For the Bald Eagle, DDT is no longer a threat since the species doesn't migrate outside the boundaries of North America. That doesn't mean the bird is safe.

Charisma is a powerful tool in conservation. If the Bald Eagle hadn't had it, hadn't been large and powerful and charismatic, would it have been chosen as a national symbol? According to legend, but never confirmed, Benjamin Franklin, believing the eagle was "of bad moral character" and unsuitable for the newly created Great Seal of the United States, preferred the turkey. But he was overruled, and the eagle, strong and majestic, was chosen. On the charisma scale, the turkey fell short.

Appeal seems to be critical to conservation. Mammals such as giant pandas, wolves, gorillas, and whales have attracted notice because of it. But how does charisma affect bird conservation? The large—the eagle, the California Condor, and the Whooping Crane; or the fast—the Peregrine Falcon; or the cute and beautiful—Piping Plovers, puffins, and bluebirds—are popular, and people are eager to endow programs to protect them. By the same token, the money provided for habitat protection of waterfowl and upland game birds depends on the willingness of hunters to conserve their quarry. But what of birds that fall outside these groups?

As was the case of the Loggerhead Shrike, ignored because it wasn't flashy, beautiful, or acrobatic, the future is far less certain for the innocent customers at our bird feeders. It is true we are more vigilant today than we were in the 1960s and 1970s. We learned the hard way. The scare of DDT shocked us into paying more attention and to bolstering protection for avian populations. Compared to mammals, birds are more difficult to keep track of. They can migrate long distances, aren't always loyal to their nesting areas, are harder to trap and mark, and represent many more species.

Citizen science has become a mainstay for bird population studies. The Cornell Lab of Ornithology and the National Audubon Society urge us all to go outside and count, to record the numbers in our notebooks, and insert them into online databases. Perhaps, and of course there is no guarantee, we will notice when a species is in trouble. We might find the cause before it is too late. The North American Breeding Bird Survey (USGS), which documents nesting birds; the Christmas Bird Count, Coastal Bird Survey, and the Great Backyard Bird Count, all sponsored by National Audubon; and Project Feeder Watch, managed by the Lab of Ornithology—these are some programs that teach participants to identify local birds and tally their numbers. Raptors are surveyed by some of these programs, but those that migrate are also counted by volunteers at sites along known migration corridors like Hawk Mountain Sanctuary in Pennsylvania. The dedication of certified bird banders has helped compile information to monitor the health and patterns of species, but with the changes in the world happening so fast, with climate change a harsh reality, we have to wonder if our efforts will be in time. The will to monitor and conserve is strong, but the question that looms before us on the eve of the fiftieth anniversary of the Bald Eagle's reintroduction, is "Will it be enough?"

Threats to wildlife persist. As one is addressed, another arises from the void. Birds of prey face poisoning from rodenticides, used by farmers and homeowners to eradicate mice, rats, and pests that attack crops and fruit trees. While this isn't meant to kill the owls and hawks that live in the area, it is lethal to them when they unwittingly eat the poisoned rodents. Other birds are victims of habitat destruction as human population increases, and development takes over open space and forestland. Chemical pesticides still used in agriculture and landscaping don't poison birds directly but contaminate the food they eat. Even wind energy projects, meant to reduce our consumption of fossil fuels, can threaten eagles and other raptors when turbines are situated in areas with high nesting activity or in flyways.

As we rest on our laurels of having protected and restored the larger-than-life eagle, we become aware of another threat that could undo all our work if we don't take action. A new enemy is lurking in our ponds and rivers. Lead, an element in ammunition used for hunting, has been found in eagle tissues, picked up as the predators scavenge animals riddled with lead shot. A study, chronicled in *Science* in 2022, examined Bald and Golden Eagles in thirty-eight states, discovering that nearly half of them had lead in their tissues. Highly toxic, banned and then reinstated as administrations changed hands, lead can weaken eagles and slow population growth. Even in states where lead has been banned for waterfowl hunting, it still lingers in the soil below the water's surface, contaminating the vegetation consumed by the ducks, which are then eaten by the eagles. The California Condor's decline was linked to lead poisoning years ago. By switching to copper and steel shot, hunters can prevent another DDT fiasco.

Because not all threats to eagles are the direct result of human activities, we need to remain steadfast in our role as sentinels at the gates of their survival. Bald Eagles can fall victim to more insidious perils as well. New discoveries have tagged diseases such as vacuolar myelinopathy (VM), found in 1994 in Bald Eagles in the southeastern states, caused by a neurotoxin that attacks the birds' brains. Produced by cyanobacterium lodged in invasive aquatic plants, the disease passes from fish, waterfowl, and other species eating the vegetation to the eagles. Research on the causes of the toxin is ongoing, including its possible link to human-related sources.

With our tendency to want to control our environment, protectors of wildlife and their habitat need to be ever vigilant. If not, Albert

Schweitzer, to whom Rachel Carson dedicated her iconic book, will be proven correct with his foreboding words, "Man has lost the capacity to foresee and forestall. He will end by destroying the earth."

With all due respect to Dr. Schweitzer, I prefer to be optimistic, to believe in the resilience of nature and to count on humankind's innate resolve to protect our earth. At my induction ceremony into the Iroquois Nation, during the "moment in the sun" for me and the wives of the chiefs, I experienced hope. For this reason, I adhere to the Thanksgiving Address of the Haudenosaunee, known as Greeting to the Natural World:

> *We put our minds together as one and thank all the Birds who move and fly about over our heads. The Creator gave them beautiful songs. Each day they remind us to enjoy and appreciate life. The Eagle was chosen to be their leader. To all the Birds—from the smallest to the largest—we send our joyful greetings and thanks. Now our minds are one.*

As my grandsons remain mesmerized by the feeding birds, I acknowledge the forces leading me from observing and working with nature to teaching about it. The memory of M3 still follows me. His courage and resilience, needed to enable his species to endure, inspires me as I strive to impart an appreciation for the natural world to others. This is to bestow a gift, for only in this way can we pass on environmental stewardship to the next generation. Today's problems, as well as those lying in wait, are surmountable, but people will have to care, to "foresee and forestall," to ensure that eagles, and all other birds, continue to return to the sky, healthy and powerful, as they were created.

My first extensive trip after Covid restrictions were lifted was to northern Maine and the Maritime provinces of New Brunswick and Nova Scotia. Larry and I visited old friends and places where he had worked in the past and we had both spent time in the 1970s. Everywhere we went people commented on the numbers of Bald Eagles they had seen over the summer, some perching on the edge of their backyard woods and some flying overhead. As one friend commented in Lubec, Maine, "As many as we now see every day, the sight never grows old." I told her about this book. Her comment was echoing in my mind as we drove out of Lubec toward Machias, passing lakes encircled by spruce trees and meadows of goldenrod, asters, and Queen Anne's lace. I surveyed the sky hoping that maybe, on this clear August day with thermals beckoning above us, we would be lucky. A large silhouette appeared high against the blue backdrop, then a flash of white behind the dark brown body, and another patch in front. My binoculars fixed on the bird, knowing I might have been imagining it, but no. Our national bird had risen to the Maine sky above me as I drove home to finish its story of survival.

# Acknowledgments

In *Return to the Sky*, any errors made in recollections of events and conversations are my fault alone. This historic project, however, was a team effort, and I wish to thank the following people, both for helping the Bald Eagle survive and for allowing its story of recovery to come to light:

The Chelsea Green staff, for all their efforts to usher this book into reality, with special gratitude to my loyal champion and editor Matthew Derr, who took a chance on my story and saw the need to bring it to the world. His love of birds and dedication to conservation allowed him to recognize its value.

Amos Eno and Nathaniel Reed of the Department of the Interior and Ogden Reid and Eugene McCaffrey of New York's Department of Environmental Conservation, who had the vision to bring back the Bald Eagle and arranged the chess pieces to make it happen.

Tom Cade, who believed in me even when I wasn't sure I believed in myself, and whose Peregrine Falcon rescue efforts set the stage for the Bald Eagle's comeback.

Jim Weaver, for easing my way and teaching me everything I needed to know about raptor biology, Bald Eagles, and hacking; and Willard Heck, Phyllis Dague, and Steve Sherrod of The Peregrine Fund for their support and encouragement.

New York's DEC staff, especially Peter Nye, Mike Allen, and all the others who worked beside me on the hillside during both 1976 and 1977 and who were there to offer valuable help whenever I needed it.

Videographers Lance Wisniewski and Pat Faust, whose video monitoring system ensured the eagles were safe and always in view.

Stanley Wiemeyer of Patuxent Wildlife Research Center and Carl Madsen of US Fish and Wildlife Service, for their donations of young eaglets to the hacking project; and Dr. Pat Redig, whose care and rehabilitation of M3 secured this special eagle a place in history.

The 1976–1977 Montezuma refuge staff: Sam Waldstein, Gene Hocutt, Vern Dewey, Ann Harrison, and their colleagues, who supported my efforts on that hillside and provided the eagles with a home from which to begin their journey to recovery.

The current staff of Montezuma, Bill Stewart and Linda Ziemba, and volunteers Jackie Bakker and Peter Saracino, for their vigilance and dedication to the Bald Eagle, which ensures the bird's survival is in good hands.

My patient and insightful editors—Ethan Gilsdorf, Dorian Fox, and Linda Spence—as well as my daughter Alix Morris and brother Michael Milburn, who painstakingly reviewed my drafts, dealt with my clichés, offered a gentle nudge when I was in the weeds, and never lost faith in this story.

Those who cheered me on throughout the project, offered advice, and gave me inspiration in so many ways: PK McClelland, Beth Johnson, Peg Rosenberry, E.B. Bartels, Carrie Koplinka, Rick Bonney, Darryl McGrath, Stephanie Schilling, David Rimmer, Kevin Porter, and Mike Waters; and my Newburyport writing community of Elizabeth Barrett, Dag Sheer, Karen Piper, Bettina Turner, Barbara Wood, Kenneth Smith, Frank Barron, Gay Cameron, Joy Sawyer-Mulligan, and Vicki Hendrickson, who encouraged me to take one essay and make it into a book.

Those iconic authors who introduced me to the magic of animals in my childhood and helped determine my future path: Thornton W. Burgess, Beatrix Potter, E.B. White, Munro Leaf, Gerald Durrell, and Joy Adamson.

My family: my parents, for instilling in me a love and respect for animals and the natural world; my siblings and children, for listening to me talk about eagles incessantly while pointing out every raptor that flew overhead; my grandchildren, for inspiring me to chronicle the past; and last but not least, Ellie, my loyal lab, who kept vigil on her couch next to my desk throughout the writing of this book.

And most of all, Larry, who has supported me and my love of animals with his usual good humor, enthusiasm, and wise counsel. I couldn't have asked for a better copilot!

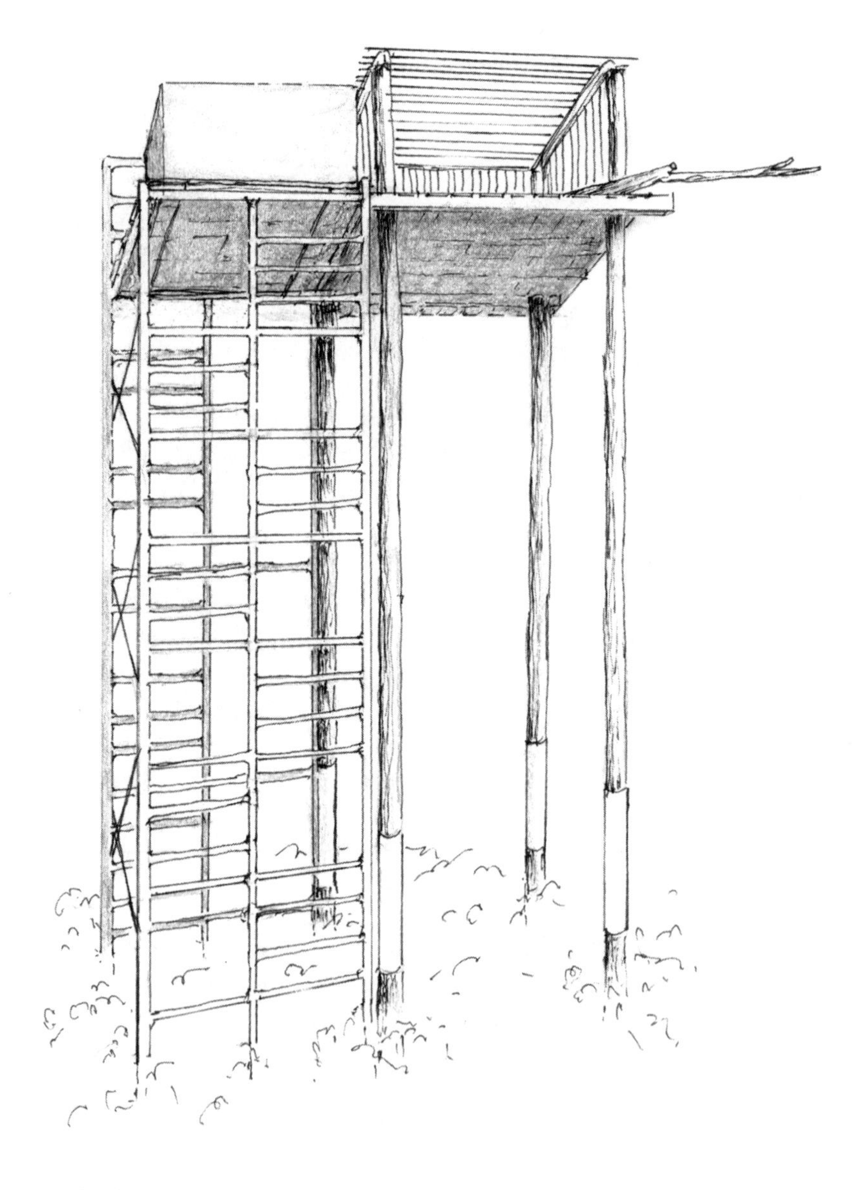

1976 hacking tower with blind and scaffolding. *Illustration by Michael Waters*

# About the Author

*Casey Morris*

Raised in a large family and surrounded by myriad orphaned creatures both domestic and wild, Tina Morris was imbued with a lifelong love of animals. After a few wrong turns and a stormy relationship with science in college, she found a way to make her life's ambition—rescuing endangered birds of prey—into a reality. Tina earned her undergraduate degree from Oberlin College and her graduate degree in ornithology and wildlife biology from Cornell University, where she helped develop the first techniques for releasing introduced Bald Eagles. Her field research ultimately became the instruction manual for eagle restoration programs in other eastern states. Tina was formally inducted as an honorary Iroquois into the Confederacy of Six Nations for her work returning the Bald Eagle to the nation's skies.